Calculus Made Even Easier

An Infinitesimal Differential Approach

By

Robert R Carter

ISBN-13: 978-1725513600
ISBN-10: 1725513609

Dedicated to my parents for their inspiration and to Betty for her encouragement and patience.

CONTENTS

INTRODUCTION

I began teaching calculus in 1982 in a public high school in south Florida and like R. Rucker (Infinity and the Mind) I too felt the frustration of teaching limits year after year to very confused students. Years later, in 1990 I read S. Thompson's delightful book Calculus Made Easy first published in 1910 (and still in print!) which has become a bellwether of "simplified" calculus books. I began a study of some old calculus texts to find out more about Thompson's simple approach. The infinitesimal differential approach actually started with one of the two founders of calculus G. Leibniz which he published in 1684. Joseph Bayma has an excellent philosophical discussion of infinitesimals in his Infinitesimal Calculus (1889). Infinitesimals were in fact used by the greatest scientists and mathematicians from D. Bernoulli to Poincare. The entire edifice of classical mechanics was built using infinitesimal differentials. Physicists and engineers continue to use them to this day.

In the 18th and 19th centuries the mathematicians were the scientists. In fact most of the advancements in mathematics came from their investigation of physics. This is hardly the case today. The 20th century began the era of specialization. Now most mathematicians rarely communicate with scientists and have developed a mathematics so abstract it is almost totally divorced from reality. Probably science and mathematics both suffer from the estrangement.

In 1734 G. Berkeley leveled serious criticism of the calculus in The Analyst. And although Newton was the target of the criticism, Leibniz infinitesimal differentials were not immune from the attack. The result of this criticism was the beginning of a reformulation of calculus using the theory of limits, which wasn't actually completed until K. Weierstrass published the modern ε - δ definition of a limit in 1854. By the end of the 19th century G. Cantor called the infinitesimal a "cholera-bacilli" that infects the calculus and infinitesimals were banished by most mathematicians as the theory of limits became the foundation and cornerstone of the calculus. In the 20th century the calculus based on limits became the "standard calculus" that is now taught in high schools and universities all over the world.

Not all mathematicians bought into this wholesale rejection of infinitesimal differentials and in 1960 the logician A. Robinson found a way to legitimize the infinitesimal using logic and set theory. This set the stage for the development of what has since become known as non-standard calculus of which there are now several types. Unfortunately most mathematicians have been slow to accept it and it is still on the fringes of mathematics.

In the 1970's two very good non-standard calculus books were published, one by J. Keisler and the other by J. Henle. These books are highly recommended to those seeking a rigorous formulation of infinitesimals rather than just an operational presentation. Another excellent book published in 1998 is A Primer of Infinitesimal Analysis by J.L Bell.

For most students in a standard calculus class, the chapter on "limits" is harder to understand than calculus itself. Students are taught left-handed limits, right-handed limits, infinite limits, indeterminate limits and more. The derivative and integral are defined as limits and all of the formulas are derived using limits. I can recall the panicked looks on student's faces when they thought that this was calculus and that they would be doing it all year long. When I explained that this was the justification of calculus and not the calculus itself they were a little relieved but some still wanted out.

The sad thing is that one can do calculus (as they did for over 150 years) without any knowledge of limits and yet limits are taught year after year to students all over the world who don't care a wit about the logical foundation of calculus. By the same token, teaching the logical and set theoretical foundations of non-standard calculus to science and engineering majors is also pointless. They have no need for it.

Eventually, in the mid 90's, I adopted the infinitesimal differential methods outlined in this book in the calculus classes that I taught with much success. With this "informal" and "operational" approach my students didn't just give up in frustration, many even showed some genuine interest.

This book then is the core of my lectures. There are no exercise sets as this is not a textbook. It is meant to be used as a supplement to gain an insight into the original concepts behind the differential notation. It is written for the non-math major who is required to take calculus and apply it in their discipline. But because of limits they are hopelessly confused about the meaning of all of the symbols, and how to apply them. The concepts presented here should promote a better operational insight into calculus.

I make no pretense of rigor. I am not a mathematician, just a simple retired high school math teacher. This is a practical approach to calculus meant for science and engineering majors. There is a small section on "derivations" that I used in class just to give students a "feeling" for the formulas. But it is far from rigorous. Again, I refer you to Keisler and Henle if it's rigor in non-standard calculus that you're after.

What I have done here is to present the differential as an infinitesimal or "ultimate part" of a variable rather than a "little bit" of a variable as Thompson did in his Calculus Made Easy because he lived in the pre-Robinson era. I hope this makes calculus even easier.

DIFFERENTIALS

INSTANTS AND DIFFERENTIALS

We will begin with a discussion of how to represent the passing of time mathematically. We will represent the passing of time as a time point T moving on a time line. If T moves from one time value t_1 to another time value t_2 we have a finite segment on the timeline called an **increment** of time and symbolized as Δt. The value of Δt is $t_2 - t_1$.

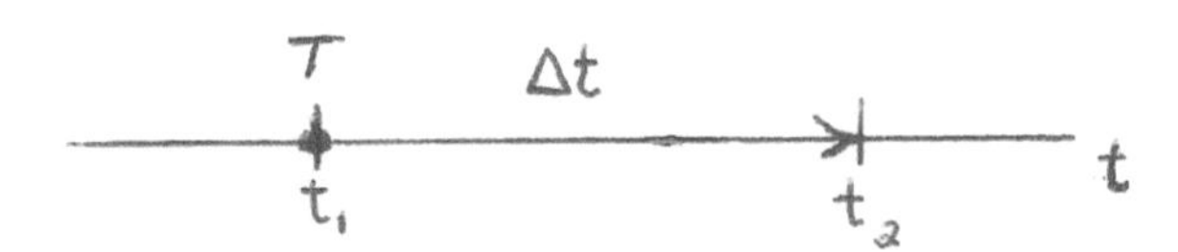

If T moves from t_1 = 3 seconds to t_2 = 8 seconds then Δt = 5 seconds and we have a 5 second increment of time. Also there are infinitely many time values between t_1 and t_2.

There are a variety of real world applications that involve instantaneous events. For these, we need an increment small enough to represent an instant. The problem is that no matter how small we make Δt, it will contain infinitely many time values. We need an increment so small that it will contain only one time value. Only an infinitely small increment will suffice because an infinitesimal increment would contain only one time value since no two time values are infinitely close.

Now since, Δt represents a finite increment of time, we will use Leibnitz symbol dt to represent an infinitesimal increment of time. Thus, dt will represent one instant because it will contain only one time value, call it t_1, since there are no other time values infinitely near t_1. The diagram below shows dt under infinite magnification (a graphic that J. Keisler started in his Infinitesimal Calculus textbook). This is also true on the micro-segment preceding t_1.

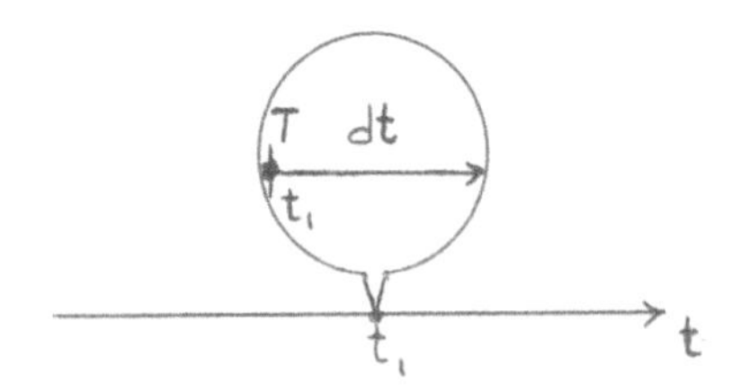

We see then that the infinitesimal increment dt can represent an instant, a "now", during which time is still passing and yet remains t_1. A point cannot represent an instant. Points have no size. How then could they have duration or add up to any real amount of time? The fleeting instant does have some duration, but it is less than any assignable value.

Students often asked, "What number does the infinitesimal dt equal?" But just as with the imaginary number i, it doesn't equal any real number. The number i because there is no real number that when squared is equal to -1 and dt because no real numbers are infinitely small. Surprisingly, even though an infinitesimal has no real number value alone, a ratio of infinitesimals can equal a real number as well as an infinite sum of them.

An infinitesimal quantity is smaller than any positive finite quantity. It is so small that its square, and all higher powers, are effectively zero. It is often referred to as a "nilsquare infinitesimal." On a timeline, the infinitesimal magnitude dt is an "ultimate lineal element," we call it a micro-segment. This micro-segment is still nevertheless a continuum so is itself divisible and therefore cannot be a point, i.e it is "nonpunctiform." Curves may even be thought of as being composed of infinitely such micro-segments.

Since all things change with time, the differential of a variable represents the infinitesimal change that occurs in that variable during an instant of time dt. So, for example, if V represents the volume of air in a balloon that is being inflated, then dV represents the infinitesimal volume pumped in an instant. Of course volume V is also related to radius r which will also change by an infinitesimal amount dr during that same instant. Once we know how to find these instantaneous changes we will then be able to determine the instantaneous *rates* of change of V and r.

Now, $d(\ )$ is called a differential operator and the variable expression inside the parentheses is the expression that we are to find the differential of. The differential of any variable u is then $d(u) = du$. This does not mean d times the variable. It means the differential of u. It is the infinitesimal or ultimate part of u. This holds for a simple variable with a power of one. If we want the differential of u^2 the answer is not du, but instead $d(u^2) = 2udu$. Note the du at the end of the expression. This is what makes the whole expression $2udu$ infinitesimal. It's infinitesimal because no matter how great the value of $2u$, the fact that it is multiplied by du makes the expression infinitely small. Of course we will also want to know $d(sin\,x)$, $d(e^x)$ and the differentials of many other expressions.

In the following pages we will examine the procedures for finding the differentials of all kinds of variable expressions. Then, we will use these differentials to find rates of change, area, volume, arc length, surface area, exponential growth/decay, and more.

MOTION ON A LINE

If a point P is moving on the x-axis then the infinitesimal distance it travels during the instant is a micro-segment dx (pictured below). We can think of a line as composed of infinitely many such micro-segments. These are the "ultimate parts" of a line not points. Points can't be the ultimate parts of a line because points have no size and can't be divided any further which violates the principle that every part of a continuum is itself a continuum and hence infinitely divisible. A point, is more like a location on a line than not an ultimate part of it. During the instant dt point P traverses the micro-segment dx.

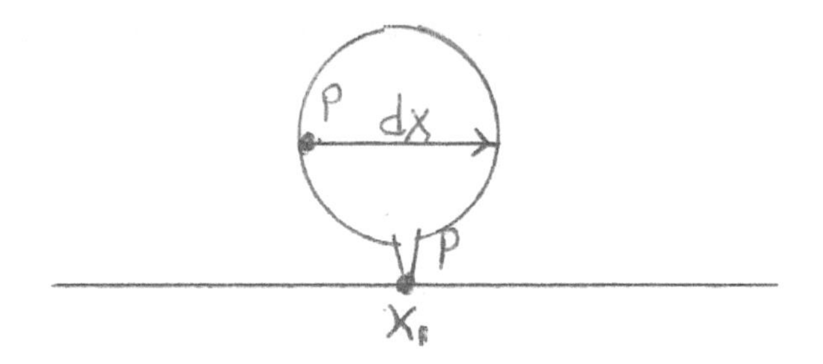

MOTION ON A CURVE

If we think of a curve as being composed of infinitely many lineal micro-segments, each of length ds, then a point P(x,y) moving on a curve from left to right will move linearly on each ds during each moment of time dt. These are the "ultimate parts" of the curve. A point has no slope but the micro segment ds that it is moving on does. The slope of ds is also the slope of the tangent line that it is a part of. Of course dy and dx are the infinitesimal changes in the rise and run of the micro-segment ds. So $\frac{dy}{dx}$ is the slope of ds of the curve which is the slope of the tangent line there, as illustrated below.

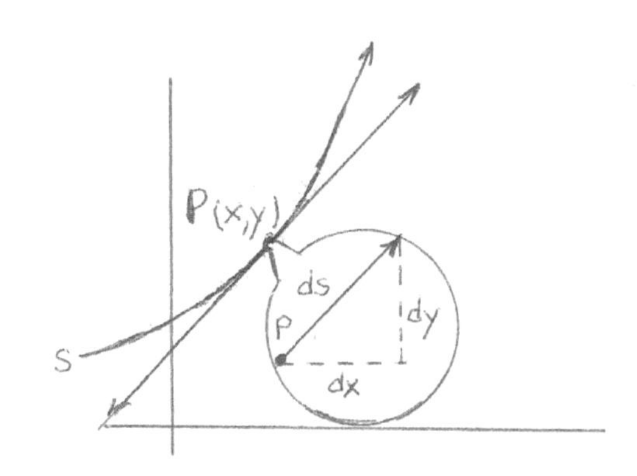

Leibniz called the triangle with sides dx, dy, and ds the "characteristic triangle." And just as there can be only one real number valued point P on a micro-segment dx of a number line (because the distance between two real number points is a finite real value not infinitesimal) there can be only one real number valued point $P(x, y)$ on a micro-segment ds of a curve.

So dx and ds are infinitesimal micro-segments of a line and a curve respectively, they are themselves infinitely divisible. They form a kind of scaffolding for the "points" (locations) on a line or curve. We call them the "ultimate parts" of lines and curves. We will see that area and volume can also be thought of as being composed of ultimate parts, dA and dV. We will also see that differentials can also be "summed up" using integrals to find the total arc length, area and volume.

Formulas

The most difficult part of every math course is the derivation or proofs of the formulas that are used. There are some 20 formulas in elementary calculus that are usually memorized. This is a daunting enough task for most beginning calculus students most of whom have little or no interest in where the formulas came from. We give them here without proof. There are some infinitesimal derivations starting on page 100. Note that all of the formulas have du multiplied at the end. The differential of a variable expression always has the differential of that variable multiplied at the end of the expression. This makes the whole expression a differential expression.

1 $d(c) = 0$ where c is a constant

2 $d(u) = du$

3 $d(c \cdot u) = c \cdot du$

4 $d(u^n) = n \cdot u^{n-1} du$

5 $d(\sin u) = \cos u \, du$

6 $d(\cos u) = - \sin u \, du$

7 $d(\tan u) = \sec^2 u \, du$

8 $d(\cot u) = - \csc^2 u \, du$

9 $d(\sec u) = \sec u \tan u \, du$

10 $d(\csc u) = - \csc u \cot u$

11 $d(sin^{-1} u) = \frac{du}{\sqrt{1-u^2}}$

12 $d(tan^{-1} u) = \frac{du}{1+u^2}$

13 $d(ln\, u) = \frac{du}{u}$

14 $d(log_a u) = \frac{du}{u \cdot ln\, a}$

15 $d(e^u) = e^u\, du$

16 $d(a^u) = a^u \cdot ln\, a\, du$

General Formulas

17 $d(u+v) = du + dv$

18 $d(u \cdot v) = udv + vdu$

19 $d(u-v) = du - dv$

20 $d\left(\frac{u}{v}\right) = \frac{vdu - udv}{v^2}$

Examples

1. $d(5x^3 + 3ln\, x) = d(5x^3) + d(3ln\, x) = 5 \cdot d(x^3) + 3 \cdot d(ln\, x) = 3 \cdot 5x^2\, dx + \frac{3dx}{x} = 15x^2 dx + \frac{3dx}{x}$

2. $d\left(\frac{1}{x^2} - tan^{-1} x\right) = d(x^{-2}) - d(tan^{-1} x) = -2x^{-3}dx - \frac{dx}{1+x^2} = \frac{-2dx}{x^3} - \frac{dx}{1+x^2}$

3. $d(x^{2/3} \cdot sin\, x) = x^{2/3} \cdot d(sin\, x) + sin\, x \cdot d(x^{2/3}) = x^{2/3} \cdot cos\, x\, dx + sin\, x \cdot \frac{2x^{-1/3}dx}{3}$

4. $d\left(\frac{sin\, x}{e^x}\right) = \frac{e^x \cdot d(sin\, x) - sin\, x \cdot d(e^x)}{(e^x)^2} = \frac{e^x \cdot cos\, xdx - sin\, x \cdot e^x dx}{e^{2x}}$

Instantaneous Rates

A ratio of differentials is an instantaneous rate of change.

$\frac{du}{dt}$ This is the “pure” rate of u i.e. the rate that u changes in an instant of time.

$\frac{du}{dv}$ This is not the “pure” rate of u, it is the rate of u compared to the rate of v because:

$$\frac{\frac{du}{dt}}{\frac{dv}{dt}} = \frac{du}{dt} \cdot \frac{dt}{dv} = \frac{du}{dv}$$

We call the pure rate of u, $\frac{du}{dt}$, the derivative of u whereas we call $\frac{du}{dv}$ the derivative of u with respect to v. As we said, this is because it is a comparison of the rate of u and the rate of v.

In standard calculus the fundamental concept is not the differential of a variable but the derivative of a function which is defined as a limit. The differential is then defined in terms of the derivative. The notion of a function and limits involves dependent and independent variables. In our approach the fundamental concept is the differential of any variable, regardless of dependence or independence, and the derivative is defined as a quotient of differentials.

When scientists use calculus, they aren't thinking in terms of functions and limits but in terms of variables representing physical quantities that are related in equations. They use differentials to find the instantaneous changes in these variable quantities and their quotients to find their instantaneous rates. Consider the equation:

$$x^2 y + xy^2 + 3x - 5y = 0$$

We can't say that y depends on x or that x depends on y. But it doesn't really matter, because differentials are determined in the same way regardless of whether a variable is dependent or independent (we will find the differentials for this equation on page 28.) Once we find the differentials we can then divide both sides of the equation by dx and obtain $\frac{dy}{dx}$ (the rate of y with respect to x) or, we can divide by dy to obtain $\frac{dx}{dy}$ (the rate of x with respect to y). Or we can even divide by dt and obtain the pure rates of both x and y i.e. $\frac{dy}{dt}$ and $\frac{dx}{dt}$.

There is a very practical notation that has been in use for a long time. If y does depend on x, we can express this as y = y(x) or simply y(x) instead of y = $f(x)$. This eliminates the need for using another "letter", f, which complicates problem solving. Note that if x depends on y, we would denote this as $x = x(y)$ or simply $x(y)$. This notation is particularly useful in science where position x, velocity v, and acceleration a depend on time t. We can denote these as $x(t)$, $v(t)$, and $a(t)$ respectively.

Just as $d()$ is a differential operator, $\frac{d}{dx}()$ is an operator for derivatives. It means find the differential of the expression in parentheses then divide by the differential dx to obtain the derivative with respect to x.

Instantaneous Velocity

The Instantaneous velocity of an object is the instantaneous rate of change of its position.

$$v(t) = \frac{dx}{dt}$$

Example 1: Find the instantaneous velocity of an object traveling on the x-axis at 4 seconds when its position is given by $x(t) = t^2 + 2t + 1$ (measured in feet)

$$v(t) = \frac{dx}{dt} = \frac{d(t^2+2t+1)}{dt} = \frac{2tdt + 2dt}{dt} = 2t + 2 = 2(2) + 2 = 6 \text{ ft/sec when } t = 2$$

Tangent Lines

The slope of a tangent line to a curve is the instantaneous rate of change of y with respect to x.

Example 1: Find the equation of the tangent line to y = x^2 at the point $P(3,9)$

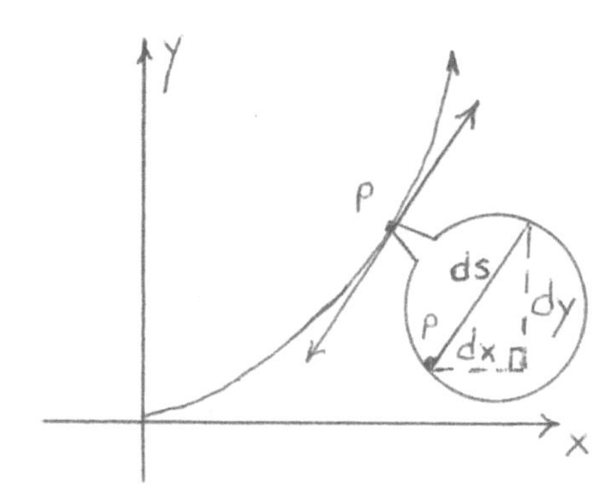

The slope of the micro segment ds containing $P(3,9)$ is $\frac{dy}{dx}$ and since y = x^2 we get $\frac{d(x^2)}{dx} = \frac{2xdx}{dx} = 2x$ and when x = 3, $2x$ = 6 which is the slope of ds.

Since the micro segment ds is on the tangent line at $P(3,9)$ the slope m of the tangent line must be 6. Using the point-slope form of a line, the tangent line equation must be y – 9 = 6(x – 3)

Example 2: Find the equation of the tangent line to the curve y = $\sin x$ at x = 0.

The slope m is $\frac{dy}{dx} = \frac{d(\sin x)}{dx} = \frac{\cos x dx}{dx} = \cos x$. Since $\cos 0 = 1 = m$ and y(0) = 0, the tangent line equation must be y – 0 = 1(x - 0).

Using Differentials

Let's consider some differential manipulations. In standard calculus manipulation of differentials is discouraged. In physics kinetic energy is labeled *T* and an expression like $dT = F\,dx$ pops up without explanation. It says that infinitesimal changes in kinetic energy occur when a force is applied over infinitesimal distances. Let's look at how it can easily be derived using differentials.

Kinetic energy $= T = \frac{1}{2}mv^2$ first, take differentials

$$d(T) = d(\tfrac{1}{2}mv^2)$$

$$dT = \frac{1}{2}m\,d(v^2) \qquad \text{mass } m \text{ is constant}$$

$$dT = 2\cdot\frac{1}{2}mv\,dv \qquad \text{Next multiply by } \frac{dt}{dt} \text{ which} = 1$$

$$dT = mv\,dv\cdot\frac{dt}{dt} \qquad \text{Next rearrange a } dt$$

$$dT = m\,\frac{dv}{dt}\,v\,dt$$

$$dT = ma\,v\,dt \qquad \text{since } \frac{dv}{dt} = a \text{ (acceleration)}$$

$$dT = F\,v\,dt \qquad \text{since } F = ma$$

$$dT = F\,\frac{dx}{dt}\,dt \qquad \text{since } v = \frac{dx}{dt}$$

$$dT = F\,dx \qquad \text{since the } dt\text{'s cancel}$$

If you are currently enrolled in a standard calculus course you're probably wondering whether these manipulations of differentials are even valid. You've been told that $\frac{dy}{dx}$ is not a fraction and can't be pulled apart. But the fact is, it can be treated as a fraction which gives scientists a powerful method for solving problems. Scientists have done it for hundreds of years. Of course in pure mathematics where analysis may involve pathological functions we can't be so caviler. But in real world problems where smooth variables reign we can.

Also, we were careful to show every step in this example because an independent differential is new to most readers, but as you adapt your thinking to differentials then you should be able to omit steps just as you would in your calculus class when working with derivatives.

Related Rates

In many instances the rate of one thing may be related to another. For example, if a balloon is being inflated then the rate that the volume is changing is related to the rate that the radius is changing. The rate of change of the volume of a pool being filled is related to the rate that the height of the water is rising. We call these related rates and we can solve them with calculus.

There is no set procedure for solving related rate problems. There are some general guidelines that can be followed though:

1) Identify with variables the things that are related.

2) If possible draw a sketch and label it with the related variables.

3) Write an equation that ties the variables together.

4) Find the differentials of the equation.

5) Substitute known values of the variables to find the requested rates with appropriate units.

Example: A rock is thrown into a pond. If the radius of the outer ripple is growing at the rate of 3 ft/sec then find the rate that the area enclosed by the ripple is growing when $r = 4\ ft$.

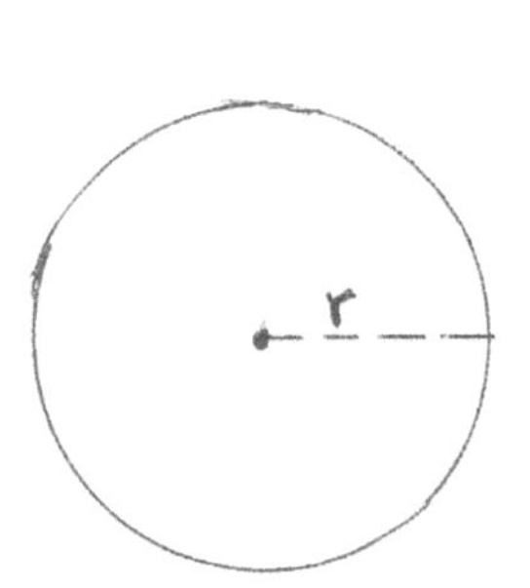

Given: $\frac{dr}{dt} = 3\ ft/sec$ Find: $\frac{dA}{dt}$ when $r = 4\ ft$

Now, $A = \pi r^2$

$$d(A) = d(\pi r^2)$$

$$dA = 2\pi r dr$$

$$\frac{dA}{dt} = \frac{2\pi r dr}{dt}$$

$$\frac{dA}{dt} = 2\pi r\frac{dr}{dt} = 2\pi(4\ ft)(3\ ft/sec) = 24\pi\ ft^2/sec$$

Example: Sand is falling onto a conical pile so that the height is always $\frac{2}{3}$ of the radius. If the rate of the height is 16 in/min then find the rate that the volume is changing when h = 6 in.

Given: $h = \frac{2}{3}r$ $\frac{dh}{dt}$ = 16 in/min

Find: $\frac{dV}{dt}$ when h = 6 in

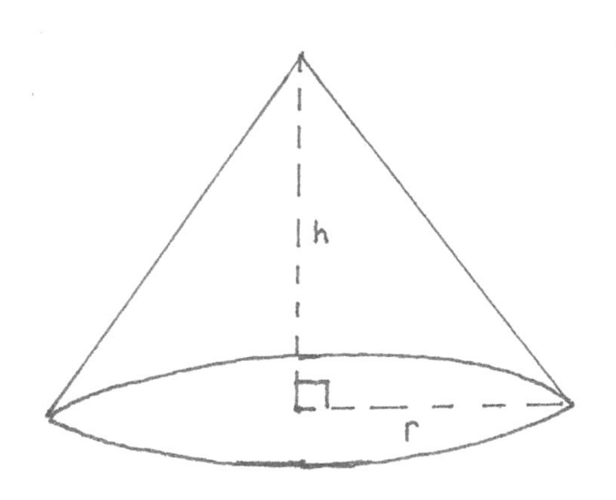

$h = \frac{2}{3}r$

when h = 6 r = 9

$dh = \frac{2}{3}\,dr$

$\frac{dh}{dt} = \frac{2}{3}\frac{dr}{dt}$

$\frac{3}{2}\frac{dh}{dt} = \frac{dr}{dt}$

$\frac{dr}{dt} = \frac{3}{2}\frac{dh}{dt}$

$\frac{dr}{dt} = \frac{3}{2}$(16 in/min)

$\frac{dr}{dt}$ = 24 in/min

$V = \frac{1}{3}\pi r^2\,h$

$V = \frac{1}{3}\pi r^2\left(\frac{2}{3}r\right)$

$V = \frac{2}{9}\pi r^3$

$d(V) = d\left(\frac{2}{9}\pi r^3\right)$

$dV = \frac{2}{9}\pi \cdot 3r^2\,dr$

$\frac{dV}{dt} = \frac{2}{3}\pi r^2\frac{dr}{dt}$

$\frac{dV}{dt} = \frac{2}{3}\pi(9\text{ in})^2(24\ in/min)$

$\frac{dV}{dt}$ = 1296π in^3/min

Example: A particle moves along the curve $y = x^2 + 3x + 1$. The rate that the x-coordinate is changing is $2 units/sec$. Find the rate that the y coordinate is changing when $x = -3\ units$

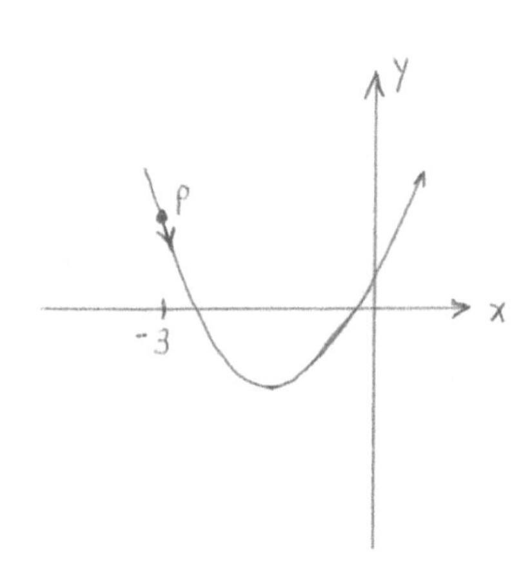

Given: $\frac{dx}{dt} = 2\ units/sec$

Find: $\frac{dy}{dt}$ when $x = -3\ units$

$y = x^2 + 3x + 1$

$dy = d(x^2 + 3x + 1)$

$dy = 2xdx + 3dx = (2x + 3)\ dx$

$\frac{dy}{dt} = (2x + 3)\frac{dx}{dt}$

$\frac{dy}{dt} = (2(-3) + 3) \cdot 2\ units/sec$

$\frac{dy}{dt} = -6\ units/sec$

The answer is negative because the particle is going down the curve.

Example: A 13' ladder is leaning against a house. When the base of the ladder is 12' from the from the house it is moving at 5 ft/sec. How fast is the ladder sliding down the wall?

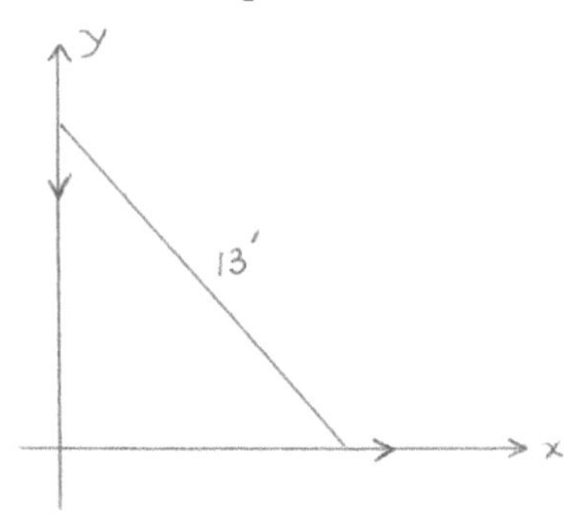

Given: $\frac{dx}{dt} = 5\ ft/sec$

Find: $\frac{dy}{dt}$ when $x = 12$

$x^2 + y^2 = 169$ note: when $x = 12$ $y = 5$

$2xdx + 2ydy = 0$

$ydy = -\ xdx$

$\frac{dy}{dt} = -\frac{x}{y}\frac{dx}{dt}$

$\frac{dy}{dt} = -\frac{12}{5} \cdot 5$

$\frac{dy}{dt} = -12\ ft/sec$

The answer is negative because the ladder is going down the wall.

DERIVATIVES

Symbolism

Leibniz tried to develop a symbolism for calculus that would accurately reflect his ideas and could be manipulated algebraically. He accomplished this so superbly with differentials that scientists have continued their use to this day. Unfortunately the meaning of his symbols have been replaced with limit concepts. Meanwhile, over the last 300 years more symbols have been added to the lexicon of calculus. Since they are widely used. On the next few pages take a look at them and how they relate to our differentials.

First let's review rates using differentials. Consider the variable expression x^2. If we want the rate of x^2 compared to x we write $\frac{d(x^2)}{dx} = \frac{2xdx}{dx} = 2x$. If we want the pure rate of x^2 with respect to time t, we write $\frac{d(x^2)}{dt} = \frac{2xdx}{dt} = 2x\frac{dx}{dt}$ which means that $2x$ is multiplied by the pure rate of x, $\frac{dx}{dt}$.

Also, if we want to represent $\frac{dy}{dx}$ evaluated at a specific point x_1, we write $\frac{dy}{dx}|_{x=x_1}$ or $\frac{dy}{dx}(x_1)$. Differentials are easy to work with. Let's relate them now to some other notations that are used.

The Capital D

The capital D with a subscript is due to L. Euler and is used to indicate that a derivative is performed on the expression in parentheses with respect to the subscript variable. So the capital D is an operator:

$$D_v(u) \text{ means find } \frac{du}{dv}$$

When working with a problem with these symbols then you can use differentials in the following way:

Example 1: $D_v(u^3) = \frac{d(u^3)}{dv} = \frac{3u^2du}{dv} = 3u^2\frac{du}{dv}$

Example 2: $D_t(\sin x) = \frac{d(\sin x)}{dt} = \frac{\cos x dx}{dt} = \cos x\frac{dx}{dt}$

Example 3: $D(x^5) = \frac{d(x^5)}{dx} = \frac{5x^4dx}{dx} = 5x^4$

Note that in example 3 there is no subscript. This means that the subscript is understood to be the same variable that is used in the parentheses.

Prime Notation

Another symbol for a derivative of y with respect to x is the prime mark ′ It is due to J.L. Lagrange. We write:

$$y' \text{ it means } \frac{dy}{dx}$$

If y depends on some other variable say t, then we put that variable inside parentheses.

$$y'(t) \quad \text{means} \quad \frac{dy}{dt}$$

Example: $y = x^2 + \sin x$ Find y'

$$y' = \frac{dy}{dx}$$

$$y' = \frac{d(x^2 + \sin x)}{dx}$$

$$y' = \frac{2xdx + \cos x\, dx}{dx}$$

$$y' = 2x + \cos x$$

Example: $y(t) = e^t + \ln t$ Find $y'(t)$

$$y'(t) = \frac{dy}{dt}$$

$$y'(t) = \frac{d(e^t + \ln t)}{dt}$$

$$y'(t) = \frac{e^t dt + \frac{1}{t} dt}{dt}$$

$$y'(t) = e^t + \frac{1}{t}$$

This notation is often very convenient.

If we have a function like $y = f(x)$, its derivative is written as, $f\,'(x)$ of course it is still true that:

$$f\,'(x) = \frac{dy}{dx} \text{ and it can also be written as } \frac{df(x)}{dx} \text{ or as } \frac{d}{dx}f(x)$$

The notation $f\,'(x)$ indicates that it is a function. We call it the "derived function."

In big D notation we can write $f\,'(x)$ as, $D_x(f(x))$ or as $D_x(\text{y})$ since $y = f(x)$.

We see then that there are several ways to write the derivative of y with respect to x they are:

$$\frac{dy}{dx}, \frac{df(x)}{dx}, \frac{d}{dx}f(x), \text{ y}', f\,'(x), D_x(\text{y}), D_x(f(x))$$

Let's revisit the dy of standard calculus. In standard calculus $\frac{dy}{dx}$ is a limit, $\lim_{\Delta x \to 0} \frac{\Delta y}{\Delta x} = \frac{dy}{dx} = f\,'(x)$. But $\frac{dy}{dx}$ is not considered to be a fraction. So, the differential dy is *defined* as $dy = f\,'(x) \cdot dx$ and dx is given a real number value. With a real value for $f\,'(x)$, dy then becomes a real value also and is used to approximate small changes in functional values, Δy, near the point. The smaller the real values for dx, the closer dy is as an approximation to the real changes in the function Δy as illustrated below.

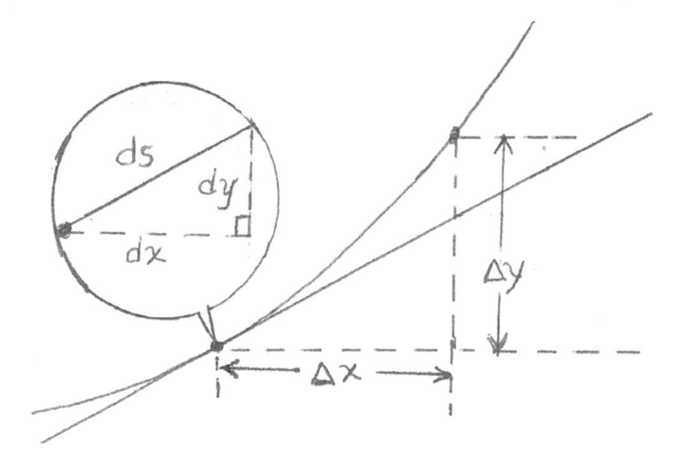

So the dy in the expression $dy = f\,'(x) \cdot dx$ is conceptually different from the dy that we have been discussing. Our dy is the infinitesimal change in y along the microsegment of the curve ds which is coincident with the tangent line at P as shown in the "bubble."

Many calculus textbooks specifically say that even though $\frac{dy}{dx}$ looks like a fraction, it isn't and should not be split apart. This of course leads to some confusion when differentials are manipulated when changing variables in integrals (p. 48)

Dot Notation

When we first discussed derivatives we said that $\frac{du}{dt}$ is the "pure" rate of u because it only involves u and time, specifically an instant of time dt. Another notation for a pure rate is the "dot" notation which began with Isaac Newton. A dot placed above a variable represents its pure rate i.e.

$$\dot{u} = \frac{du}{dt}$$

Example: $u = 3t^2 + 2t - 1$

$$\dot{u} = \frac{du}{dt}$$

$$\dot{u} = \frac{d(3t^2 + 2t - 1)}{dt}$$

$$\dot{u} = \frac{6tdt + 2dt}{dt}$$

$$\dot{u} = 6t + 2$$

Velocity

Previously, we stated that if an object is in motion and its x-position is expressed with respect to time t, then the instantaneous rate of its change in position is $\frac{dx}{dt}$ which is its instantaneous velocity v. That is, v is the pure rate that its x-position changes so $\dot{x} = v$.

Example: A particle moves on the x-axis so that its x-position is given by $x = e^t + t$. Find its velocity formula.

$$v(t) = \dot{x} = \frac{dx}{dt} = \frac{d(e^t + t)}{dt} = \frac{e^t dt + dt}{dt} = e^t + 1$$

The Chain Rule

The chain rule in standard calculus seems much more complicated than it really is. The chain rule is used on composite functions. That is, if $u = u(v)$ and $v = v(w)$ then $u = u\big(v(w)\big)$. So u is a function of v and v is a function of w. With the differential approach the chain rule is built right into it making it unnecessary to use the chain rule.

The chain rule: $\frac{du}{dw} = \frac{du}{dv} \cdot \frac{dv}{dw}$

This is clearly true as the dv's would cancel. For example 1 we show the direct application of the chain rule:

Example 1): Let $u = v^4$ and $v = w^3$ Find $\frac{du}{dw}$

$\frac{du}{dv} = 4v^3$ $\frac{dv}{dw} = 3w^2$ So, $\frac{du}{dw} = 4v^3 \cdot 3w^2 = 4(w^3)^3 \cdot 3w^2 = 4w^9 \cdot 3w^2 = 12x^{11}$

Now let's use our usual differential approach that has the chain rule built in:

$$\frac{du}{dw} = \frac{d(v^4)}{dw} = \frac{4v^3dv}{dw} = \frac{4v^3 \cdot 3\,w^2dw}{dw} = 4v^3 \cdot 3w^2 = 4(w^3)^3 \cdot 3w^2 = 12w^{11}$$

Notice the dv and dw pop out in the numerator and the dw cancels at the end.

Finally we can also just substitute at the beginning of the problem and then differentiate:

Since $u = v^4 = (w^3)^4 = w^{12}$ we have, $\frac{du}{dw} = \frac{d(w^{12})}{dw} = \frac{12w^{11}dw}{dw} = 12w^{11}$

Example 2) Let $u = v^2$ and $v = \cos w$ find $\frac{du}{dw}$.

$$\frac{du}{dw} = \frac{d(v^2)}{dw} = \frac{2vdv}{dw} = \frac{2\cos w(-\sin w)dw}{dw} = \frac{-2\sin w \cdot \cos w dw}{dw} = -2\sin w \cos w$$

Second Derivatives

The derivative of a derivative is called a second derivative. Simply put it's just the derivative performed twice in succession. The symbols in the different notations are:

In differential notation the second derivative is: $\frac{d\left(\frac{dy}{dx}\right)}{dx}$ or $\frac{d}{dx}\left(\frac{dy}{dx}\right)$. It's also symbolized as $\frac{d^2y}{dx^2}$

In big D notation: $D_x{}^2(y) = D_x(D_x(y))$

In prime notation: $y'' = (y')'$

Note: The expression $\frac{d^2y}{dx^2}$ is not a "true" fraction because it is not the actual differential expression for $y''(x)$. It should not therefore be pulled apart like $\frac{dy}{dx}$. It is in fact an unfortunate symbolism that is based on the expression $\frac{d}{dx}\left(\frac{dy}{dx}\right)$ in which it "looks like" the d's and the dx's are to be multiplied i.e. "squared." They are not. Technically the second derivative should be based on $\frac{d\left(\frac{dy}{dx}\right)}{dx}$. I refer readers to Appendix B of Jonathan Bartlett's Calculus From The Ground Up for an explanation of what the true differential expression is for the second derivative and how it's derived from $\frac{d\left(\frac{dy}{dx}\right)}{dx}$.

The following examples illustrate the second derivative of $\sin x$ in all three notations:

1) $\frac{d^2}{dx^2}(\sin x) = \frac{d}{dx}\left(\frac{d\,(\sin x)}{dx}\right) = \frac{d}{dx}\left(\frac{\cos x\, dx}{dx}\right) = \frac{d}{dx}(\cos x) = \frac{-\sin x\, dx}{dx} = -\sin x$

2) $D^2(\sin x) = D(D(\sin x)) = D(\cos x) = -\sin x$

3) $y = \sin x$ and $y' = \cos x$ so, $(y')' = (\cos x)' = -\sin x$

Acceleration

If an object is in motion with velocity v, then the rate of change of its velocity is its acceleration, a

$a = \frac{dv}{dt} = \dot{v}$. And since $v = \dot{x}$, $\dot{v} = \ddot{x}$. So a is the second derivative of position.

Example: Let $x = t^2$ then, $v = \dot{x} = \frac{dx}{dt} = \frac{d(t^2)}{dt} = \frac{2tdt}{dt} = 2t$ and $a = \dot{v} = \frac{dv}{dt} = \frac{d(2t)}{dt} = \frac{2dt}{dt} = 2$

Implicit Functions

If y is expressed in terms of x as in the following equation:

$$y = x^3 + x^2 + 1$$

then we can say that y is an explicit function of x i.e. $y = f(x)$ or $y(x)$ and we can write:

$$f(x) = x^3 + x^2 + 1$$

In standard calculus this is the preferred type of equation because the derivative $f\,'(x)$ is defined as the limit of an explicit function as its increments approach zero. In science real world problems are more likely the kind that mathematicians call "implicit" functions. An implicit function is like the following:

$$x^2y + xy^2 + 3x - 5y = 0$$

In this equation mathematicians say that it is "implied" that y is a function of x even though it's not explicitly shown. In standard calculus "implicit differentiation" is used to find $\frac{dy}{dx}$. We will show this on p. 29

But first we use our differential approach on the equation then divide by dx to find $\frac{dy}{dx}$.

$$d(x^2y + xy^2 + 3x - 5y) = d(0)$$

$$d(x^2y) + d(xy^2) + d(3x) - d(5y) = 0$$

$$x^2d(y) + y\,d(x^2) + xd(y^2) + y^2d(x) + 3d(x) - 5\,d(y) = 0$$

$$x^2dy + y \cdot 2x\,dx + x \cdot 2y\,dy + y^2dx + 3\,dx - 5\,dy = 0$$

$$x^2dy + 2xy\,dy - 5\,dy + 2xydx + y^2dx + 3\,dx = 0$$

$$(2xy + y^2 + 3)\,dx + (x^2 + 2xy - 5)\,dy = 0$$

$$(x^2 + 2xy - 5)\,dy = -\,(2xy + y^2 + 3)\,dx$$

$$\frac{dy}{dx} = \frac{-\,(2xy + y^2 + 3)}{(x^2 + 2xy - 5)}$$

In standard calculus the derivative is the fundamental concept so to find the derivative of an implicit function we differentiate "implicitly" by applying the chain rule to any expression with y. So for example, the derivative of x with respect to x is $\frac{dx}{dx} = D_x(x) = 1$ but the derivative of y with respect to

x is not 1, $\frac{dy}{dx}$ = $D_x(y)$ = y '. The derivative of x^2 is $D_x(x^2)$ = $2x$ but the derivative of y^2 is not 2y. In fact, $D_x(y^2)$ = 2yy ' because y is "implicitly" dependent on x so the chain rule must be applied to the y so we get get y 'at the end. We now show the implicit differentiation of standard calculus on the same equation above to find y '. We will use the $D_x()$ operator now.

$$x^2y + xy^2 + 3x - 5y = 0$$

$$D_x(x^2y + xy^2 + 3x - 5y) = D_x(0)$$

$$D_x(x^2y) + D_x(xy^2) + D_x(3x) - D_x(5y) = 0$$

$$x^2D_x(y) + yD_x(x^2) + xD_x(y^2) + y^2D_x(x) + 3D_x(x) - 5D_x(y) = 0$$

$$x^2y' + y \cdot 2x + x \cdot 2yy' + y^2 \cdot 1 + 3 \cdot 1 - 5 \cdot y' = 0$$

$$x^2y' + 2xy + 2xyy' + y^2 + 3 - 5y' + 0$$

$$x^2y' + 2xyy' - 5y' + 2xy + y^2 + 3 = 0$$

$$y'(x^2 + 2xy - 5) + (2xy + y^2 + 3) = 0$$

$$y'(x^2 + 2xy - 5) = -(2xy + y^2 + 3)$$

$$y' = \frac{-(2xy + y^2 + 3)}{(x^2 + 2xy - 5)}$$

Clearly, this procedure is a bit trickier than just using differentials.

There is yet another method of finding $\frac{dy}{dx}$ of $f(x, y) = 0$ which is even easier than differentials, but it involves the concept of a **total differential**. We will show this next.

Total Differentials

To find $\frac{dy}{dx}$ of our implicit function $f(x, y) = x^2y + xy^2 + 3x - 5y$ of the previous examples, we will use what's called the total differential of f, df. This is defined as:

$$df = \frac{\partial f}{\partial x}dx + \frac{\partial f}{\partial y}dy$$

$\frac{\partial f}{\partial x}$ and $\frac{\partial f}{\partial y}$ are called partial derivatives.

$\frac{\partial f}{\partial x}$ means that we must find the derivative of f while treating y as if it were a constant. So:

$$\frac{\partial f}{\partial x} = 2xy + y^2 + 3$$

$\frac{\partial f}{\partial y}$ means find the derivative of f while treating x as if it were a constant. So:

$$\frac{\partial f}{\partial y} = x^2 + 2xy - 5$$

So for $f(x, y) = x^2y + xy^2 + 3x - 5y = 0$ the total differential method of finding $\frac{dy}{dx}$ is:

$$d(x^2y + xy^2 + 3x - 5y) = d(0)$$

$$\frac{\partial f}{\partial x}dx + \frac{\partial f}{\partial y}dy = 0$$

$$(2xy + y^2 + 3)\,dx + (x^2 + 2xy - 5)\,dy = 0$$

$$\frac{dy}{dx} = \frac{-(2xy + y^2 + 3)}{(x^2 + 2xy - 5)}$$

Note that $\frac{-(2xy + y^2 + 3)}{(x^2 + 2xy - 5)}$ is actually just $-\frac{\frac{\partial f}{\partial x}}{\frac{\partial f}{\partial y}}$

So, given an implicit function $f(x, y) = 0$ a short cut to find $\frac{dy}{dx}$ would be:

$$\frac{dy}{dx} = -\frac{\frac{\partial f}{\partial x}}{\frac{\partial f}{\partial y}}$$

Summary

In this chapter we have discussed derivatives paying particular attention to the different notations we have for them and how they are related to differentials. Unfortunately many students are never shown these simple concepts and are consequently befuddled by all of the different symbols. Any student entering the sciences needs an efficient way of "doing" calculus and an understanding of the different symbols at their disposal. The "rigorous" theoretical limit approach to calculus should be left to the math majors.

Most students' sense that there must be a reason that $\frac{dy}{dx}$ looks like a fraction. We have stated that it is a fraction. Telling students that it's not just cloaks differentials in unnecessary mystery and obscures their usefulness. But a word of warning, even though $\frac{d^2y}{dx^2}$ also looks like a fraction it's not because it's not the actual differential quotient for the second derivative (page 27). It is just the symbol that is used for the second derivative. It should therefore not be separated as we do with $\frac{dy}{dx}$.

One last note about the differential dy. We said that in standard calculus dy represents a real number value but in our approach it is an infinitesimal. We also said that $\frac{dy}{dx}$ is a fraction, we can therefore write $dy = \frac{dy}{dx} \cdot dx$ and we can say then that the derivative $\frac{dy}{dx}$ is the direct proportionality factor between the rise dy and the run dx of the microsegment ds of a curve. In fact it was called the differential coefficient of dx beginning in the eighteenth century.

CURVE SKETCHING

Derivatives and curves

Max/Min Points

We can think of a curve as being traced out by a point moving from left to right. Since it moves linearly during each instant it creates infinitely many micro-segments we call ds from moment to moment. We stated that the slope of each ds is $\frac{dy}{dx}$ = y ′. So if y ′ is positive the slope of ds is positive which means that it is slanted up / so the graph is rising. If y ′ is negative the slope of ds is negative and it is slanted down \ so the graph is falling.

The point at which a curve changes from rising to falling we call a local maximum point. Here the derivative changes sign from positive to negative The point at which the curve changes from falling to rising we call a local minimum point. Here the derivative changes sign from negative to positive. Since the curve changes sign at a local max/min point then its slope there must be either zero or undefined there. If the slope is zero then the max/min point must be on a horizontal ds, a "hump." If the slope is undefined then the max/min point must be a corner, cusp, or knot.

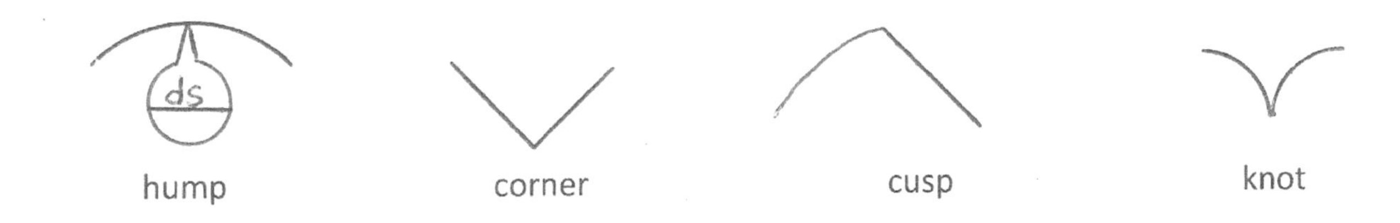

If a curve is rising then stops abruptly the point at which it stops is called an end-point max. If it falls and stops, its stopping point is called an end-point minimum. Unless it's undefined then no max/min.

Critical Values

Any values at which y ′ is zero (or undefined) are called a critical values. Every max/min point occurs at a critical value. But not all critical values are max/min points, they could be points of inflection, vertical ds, or discontinuities. To find max/min points we must check around all critical values. We do this by setting the derivative equal to zero and solving for x. We then plot these values and the undefined values on an x line and check the sign of y ′ to the left and right of them to see if there is a sign change there.

Example: Given $y = \frac{x^3}{3} + \frac{x^2}{2} - 6x + 1$ find where y is rising and falling and all max/min points.

$$y' = \frac{dy}{dx} = \frac{d\left(\frac{x^3}{3} + \frac{x^2}{2} - 6x + 1\right)}{dx} = \frac{x^2dx + xdx - 6dx}{dx} = x^2 + x - 6 = (x + 3)(x - 2) = 0 \text{ when } x = -3, 2$$

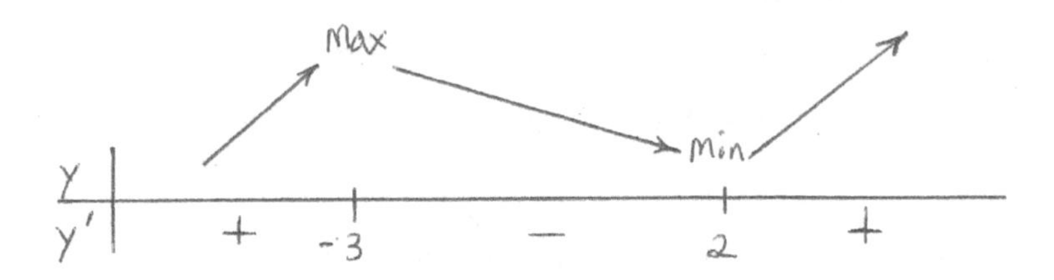

Note that when x = -3 or x = 2 we have a sign change from positive to negative at x = -3 so we have a max there. It changes from negative to positive at x = 2 so we have a min there. If we want to know the y values of these max/min points then we simply evaluate y(-3) and y(2).

Concavity

The concavity of a graph is determined by the second derivative. Intuitively concavity is whether a graph is "cupped" up (concave) or down (convex). Since the ultimate parts of a curve are straight lineal segments ds, it is the sequence of linelets that produce this effect. If the second derivative is positive then the graph is concave, if negative then it's convex.

concave convex

The location at which the concavity changes (if it does) is called "a point of inflection" or POI. These only occur at the critical values of the second derivative (as with max/min values of the first derivative)

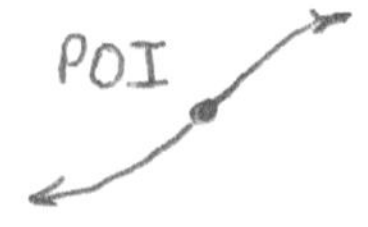

Example: Given that y = $\frac{x^3}{3} + \frac{3x^2}{2} + 2x + 1$ find where the graph is concave and convex and POI.

We need the sign of the second derivative. So differentiate twice and set the second derivative equal to zero.

$$y' = \frac{dy}{dx} = \frac{d\left(\frac{x^3}{3} + \frac{3x^2}{2} + 2x + 1\right)}{dx} = \frac{x^2dx + 3xdx + 2dx}{dx} = x^2 + 3x + 2$$

$$y'' = \frac{dy'}{dx} = \frac{d(x^2 + 3x + 2)}{dx} = \frac{2xdx+3dx}{dx} = 2x + 3 = 0 \quad \text{when } x = -\frac{3}{2} \text{ which is a POI.}$$

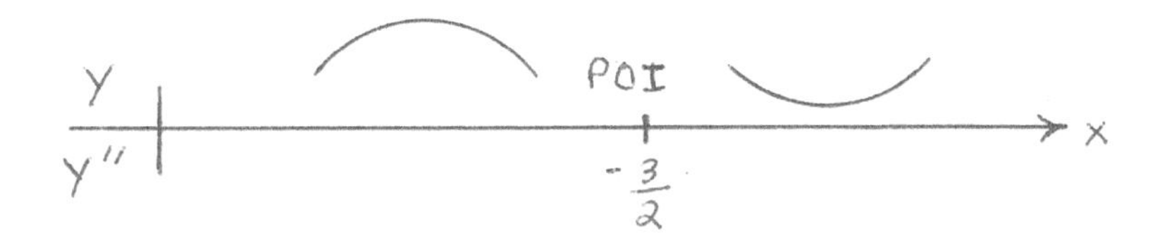

Note that if $x < -\frac{3}{2}$ then y'' < 0 so the graph is convex to the left and if $x > -\frac{3}{2}$ then y'' > 0 so graph is concave to the right of $x = -\frac{3}{2}$. Since the concavity changes at $x = -\frac{3}{2}$ we have a POI there.

Continuity/Discontinuity

In terms of infinitesimals, we say that a function f is continuous at a point "a" if whenever x is infinitely close to a, then $f(x)$ is infinitely close to $f(a)$. If this is true for all "a" in its domain then it is a continuous function and it's graph has no holes, jumps, asymptotes, or wild oscillations. Intuitively, this means that it can be drawn without lifting pencil from paper. Otherwise it is discontinuous. This is important because derivatives don't exist at discontinuities. I refer interested readers to Keisler or Henle for a more detailed infinitesimal analysis of continuity.

It was believed until the 19th century that if a function was continuous it must have derivatives (except possibly at isolated points like corners or cusps). But in 1872 Weierstrass published the first example of a function that was continuous everywhere in its domain but was differentiable nowhere. This was very disturbing to classical scientists but fortunately these types of functions did not show up in classical physics. In fact they were labeled "pathological functions."

Derivatives and Graphs

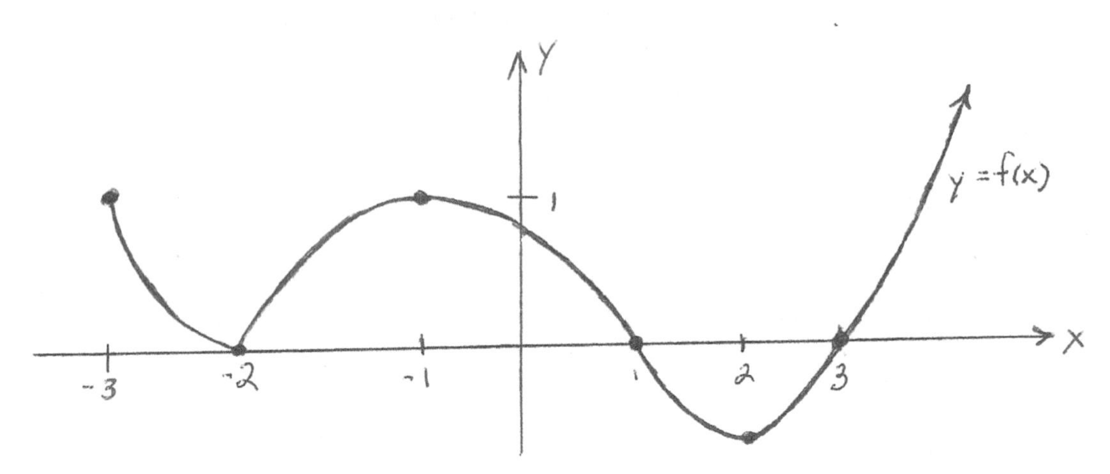

1) Where is $f\,'$ positive?

2) What is the endpoint max?

3) Where is $f\,'$ negative?

4) Where is $f\,' = 0$

5) Where is $f\,'$ undefined?

6) What are the coordinates of the local max?

7) What are the coordinates of the P.O.I.?

8) Where is f'' positive?

9) What are the critical values?

10) What are the coordinates of the local min?

Answers on the next page.

1) Where is f' positive? For $-2 < x < -1$ and $2 < x < \infty$

2) What is the endpoint max? $y = 1$

3) Where is f' negative? For $-3 < x < -2$ and $-1 < x < 2$

4) Where is $f' = 0$ At $x = -1$ and $x = 2$

5) Where is f' undefined? At $x = -2$

6) What are the coordinates of the local max? (-1,1)

7) What are the coordinates of the P.O.I.? (-2,0) and (1,0)

8) Where is f'' positive? For $-3 < x < -2$ and $1 < x < \infty$

9) What are the critical values? $x = -2, -1, 2$

10) What are the coordinates of the local min? (-2,0) and (2,-1)

Summary

We can of course combine all of these concepts to find out all about the graph of a function; where it's rising/falling, max/min points, convex/concave, and points of inflection. In fact this was done for 300 years. With today's technology no one would go through this trouble except a beginning calculus student for an assignment. Of course an understanding of what the derivatives are doing when looking at a graph, in my opinion, should deepen a student's grasp of calculus.

The following is a summary of the concepts discussed about derivatives and graphs:

1) On intervals where the graph is rising the 1st derivative is positive, falling it's negative.

2) We know that at "humps" whether max or min, the 1st derivative is zero. At corners, cusps, or knots the derivative is undefined. And all of these points are called critical points.

3) We know that on intervals where the 2nd derivative is positive the graph is concave, negative it's convex. The point where concavity changes is a point of inflection.

4) The absolute max is the highest point of the graph, the absolute min is the lowest. Graphs may have neither.

5) A local or relative max/min point is a hump, corner, cusp, or knot. All of these points have graph on the left and right sides of them, "leading right up to them."

6) An end-point max/min point has other points of the graph on only one side either to the right or left of them.

OPTIMIZATION

Optimization

In the previous section we discussed max/min values on a graph. We can extend this concept to max/min values of processes that occur in the real world. These are called optimization problems. Whether it is a maximum velocity, profit, stress load, or dispersal rate or if we need the minimum cost expenditure, pressure, or energy output we use the derivative to "optimize." Generally when solving such problems we:

1) Pick "letters" to represent the variables of the problem, like a for area or w for weight. Don't just use x , y, and z.

2) Write a primary equation that ties the variables together.

3) Write a secondary equation that can be used for substitution purposes in the primary equation.

4) Find differentials of both equations and substitute differentials of the secondary into the primary equation.

5) Divide by the appropriate differential to obtain a derivative. Set the derivative equal to zero and locate the critical values.

Example 1) A rectangle is to have the smallest perimeter possible, but its area must be 16 in^2. Find the dimensions of the rectangle.

Let l be the length, w the width, and P the perimeter.

Primary equation: $P = 2l + 2w$

Secondary equation: $l \cdot w = 16$

Differentiate the equations: $dP = 2dl + 2dw$

$$ldw + wdl = 0 \Rightarrow ldw = -wdl \Rightarrow dw = -\frac{w}{l}dl$$

Since we want to minimize P we substitute $-\frac{w}{l}dl$ into the primary differential equation for dw and set it equal to zero then solve.

$$dP = 2dl + 2dw$$

$$dP = 2dl + -\frac{2w}{l}dl$$

$$dP = \frac{2l}{l}\,dl - \frac{2w}{l}\,dl$$

$$dP = 2dl + -\frac{2w}{l}\,dl$$

$$dP = \frac{2l}{l}\,dl - \frac{2w}{l}\,dl$$

$$\frac{dP}{dl} = \frac{2l - 2w}{l}$$

Set $2l - 2w = 0 \Rightarrow l = w$

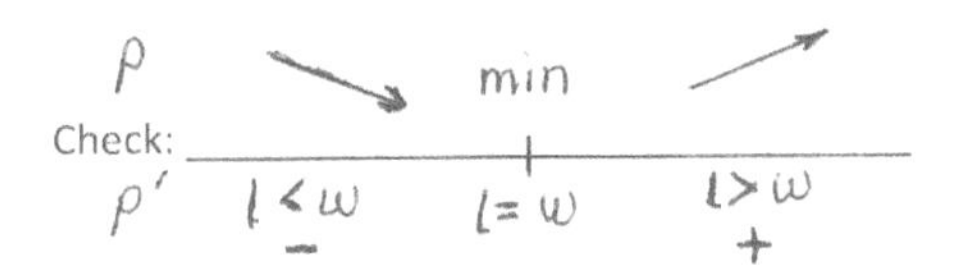

So the rectangle is a 4 in X 4 in square.

Example 2): A rectangle has its base on the $x\text{–}axis$ and its upper vertices on the parabola $y = 12 - x^2$. What is the largest area the rectangle can have?

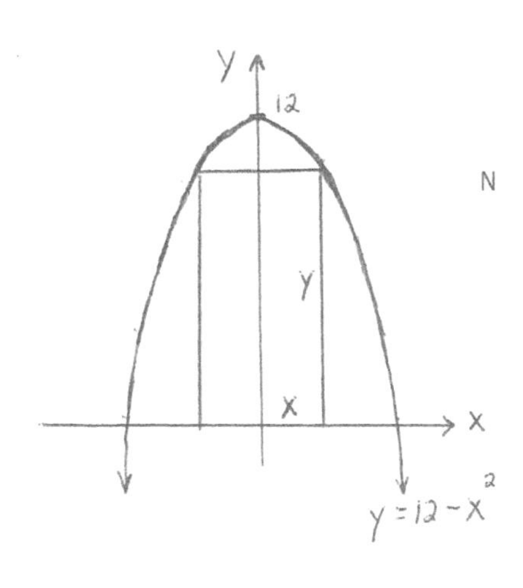

N

Primary equation: $A = 2xy$ Secondary equation: $y = 12 - x^2$

$$dA = 2(xdy + ydx)$$

$$dy = -\,2x\,dx$$

$$dA = 2(x(-\,2x\,dx) + ydx)$$

$$dA = 2(y - 2x^2)\,dx$$

$$\frac{dA}{dx} = 2(12 - x^2 - 2x^2)$$

Set $12 - x^2 - 2x^2 = 0$

$$3x^2 = 12$$

and $x^2 = 4 \Rightarrow x = 2 \Rightarrow y = 8$

A
max
Check:
A'
+
2
–

So the largest area is $A = 2xy = 2(2)(8) = 32$

INTEGRATION

Integration

The integral $\int$ is simply the inverse of d. Since differentiation and integration are inverse operations, they cancel like inverse functions $f^{-1}(f(x)) = x$ and $f(f^{-1}(x)) = x$ So,

$$\int d(u) = u \quad \text{and} \quad d\int u\,dx = u\,dx$$

Example: $\int d(\sin x) = \sin x$ and $d\int \sin x\,dx = \sin x\,dx$

We rarely have this situation though. In almost all cases the integral sign isn't followed straight away by the differential d but by a variable expression with a differential at the end like $\int 3x^2 dx$.

In this case we have to find the expression that $3x^2 dx$ is the differential of. Now, we know that $d(x^3) = 3x^2 dx$ so since

$$\int d(x^3) = x^3 \text{ the solution of } \int 3x^2 dx \text{ must also be } x^3.$$

But recall that when finding the differential of an expression any "dangling" constant c disappears because $d(c) = 0$ i.e. $d(x^3 + c) = d(x^3) + d(c) = 3x^2 dx$. So when integrating, the "complete" answer should contain a constant c that might have been in the original expression.

$$\int 3x^2 dx = x^3 + c$$

Because constants "disappear" when differentiating they should "re-appear" when integrating. We call this an indefinite integral. We can integrate some integrals by simple inspection.

$$\int \cos x\,dx = \sin x + c \qquad \int e^x dx = e^x + c \qquad \int \frac{dx}{1+x^2} = \tan^{-1} x + c$$

But most integrals will require a "method" to solve them. If the given information (called initial conditions) make it possible to determine the value of c we then have a "particular integral."

Example: Find $\int 2xdx$ given that when $x = 3$ the integral is 11

$$\int 2xdx = x^2 + c$$

$$11 = 3^2 + c$$

$$2 = c$$

$$\text{So } \int 2xdx = x^2 + 2$$

$$\text{In general then: } \int f(x)dx = F(x) + c$$

$$\text{Where } d(F(x) + c) = f(x)dx$$

We also call $F(x)$ the "antiderivative" of $f(x)$ since $\frac{dF}{dx} = f(x)$

There are two basic properties of integrals we can use right away:
1) Just as a differential distributes across a sum or difference $d(u + v) = du + dx$ so does an integral and the dx

$$\int(u + v)\, dx = \int udx + \int vdx$$

2) Just as constant coefficients move across a differential $d(cu) = cd(u)$ they move across integrals too.

$$\int cudx = c \int udx$$

The expression following an integral sign must have the same variable that is in the differential. So if we have $\int udx$ the u must be expressed in terms of x (or the dx in terms of du).

Example: $\int udx$ cannot be integrated. But if $u = 3x^2$ then $\int udx = \int 3x^2 dx = x^3 + c$. If we try to keep u and replace dx we have $du = 6xdx$ then $dx = \frac{du}{6x} = \frac{du}{6\frac{\sqrt{u}}{\sqrt{3}}}$ and the integral $\int udx = \int u \frac{du}{6\frac{\sqrt{u}}{\sqrt{3}}}$ but clearly this would be far more laborious to simplify and integrate!

Note that the derivative of an integral is: $\frac{d}{dx}\int u\, dx = d \int u = u$

Example: $\frac{d}{dx}\int \sin x\, dx = d \int \sin x = \sin x$

One final note, the integral sign $\int$ actually looks like an elongated "S." That's because that was Leibniz intention. It represents a sum that's actually solved as an inverse. He said that an integral "summed up" the infinity of differential parts of a variable and gives the "whole" variable. The realization that an integral "sum" is actually solved as the "inverse" of the differential was one of the most important discoveries of the modern age. In the 18th century it led to a virtual explosion in classical physics.

INTEGRATION METHODS

The Power Formula

$$\text{Since } d\left(\frac{x^{n+1}}{n+1} + \text{c}\right) = x^n\, dx$$

$$\int d\left(\frac{x^{n+1}}{n+1} + \text{c}\right) = \int x^n\, dx$$

$$\frac{x^{n+1}}{n+1} + \text{c} = \int x^n\, dx$$

$$\text{So, } \int x^n\, dx = \frac{x^{n+1}}{n+1} + \text{c}$$

This is called the power formula for integrals. This shows that the integral of a variable to a power (it actually applies to negative and fractional powers as well) is the variable to a power that is one more than the original power over the new power with a constant of integration.

Examples:

1) $\int x^5\, dx = \frac{x^6}{6} + \text{c}$

2) $\int x^{-5} dx = \frac{x^{-4}}{-4} + \text{c}$

3) $\int x^{2/3} dx = \frac{x^{5/3}}{5/3} + \text{c} = \frac{3}{5} x^{5/3} + \text{c}$

4) $\int x^{-2/3} dx = \frac{x^{1/3}}{1/3} + \text{c} = 3x^{1/3} + \text{c}$

The Substitution Method

Some integrals can be solved by simple recognition, while other seemingly simple integrals are unsolvable. Some integrals can be solved with closed form expressions (a finite number of expressions) while others have only an infinite series expansion (usually learned in Calculus II). The power of our differential approach will be seen in the various substitution methods in the following pages.

There are calculators now that can solve indefinite integrals. They even give formulaic solutions, so the days of memorizing integration methods are numbered. It doesn't hurt however (at least not much) to develop more skill at manipulating differentials algebraically because this is where proficiency with differentials is a real asset. Furthermore one still needs to know how to set the integral up.

The substitution method involves the power rule and a u-substitution. It involves deciding on an expression to be u and then checking to see if du is present and if it is we use the power rule or simple recognition to solve it.

Recall that the power rule for integrals is:

$$\int u^n du = \frac{u^{n+1}}{n+1} + \text{c}$$

This applies regardless of what u is.

Example 1) Suppose we have the integral $\int (x+1)^5 dx$. This would be a very tedious problem to expand and integrate term by term. But if we let $u = (x+1)$ we see that $du = d(x+1) = d(x) + d(1) = dx$ so substituting this into the integral we get $\int u^5 du$. The solution then is $\frac{u^6}{6} + \text{c} = \frac{(x+1)^6}{6} + \text{c}$.

Example 2)

$\int (x^2+1)^4 2xdx$ In this example $u = (x^2+1)$. Note that $du = d(x^2+1) = d(x^2) + d(1) = 2xdx$ so substituting we get $\int u^4 du$. The solution is $\frac{u^5}{5} + \text{c} = \frac{(x^2+1)^5}{5} + c$. Students often ask "What happened to the $2xdx$?" Well, nothing really, because $2xdx$ is actually du and a differential never appears in the answer of an integral whether its du or $2xdx$ because integration "inverses" the differential to the variable expression it came from. So there is no differential in the answer.

Example 3)

$\int \cos 5xdx$. In this example, we let $u = 5x$ then $du = 5dx$. We see that we need a 5 before the dx then we can replace it with du. This is easily handled by multiplying the integral by 1 written as $\frac{1}{5} \cdot 5$ So we put the 5 in front of the dx and move the $\frac{1}{5}$ across the integral (this can only be done with constants NOT variables) and we get: $\frac{1}{5}\int \cos 5x \cdot 5dx$ substituting back into the original problem we get $\frac{1}{5}\int \cos u \cdot du = \frac{1}{5}\sin u + \text{c} = \frac{1}{5}\sin 5x + \text{c}$

Integration Using Ln

If an integral has a fraction in which the differential of the denominator is found in the numerator we then can use the integral formula for the natural log function to solve it:

$$\int \frac{du}{u} = ln\ |u| + \text{c}$$

Example: $\int \frac{2xdx}{x^2+1}$ in this example $u = x^2 + 1$ so $du = d(x^2 + 1) = d(x^2) + d(1) = 2xdx$. Substituting we get $\int \frac{du}{u}$. The solution is $\ln |u| + \text{c} = \ln |x^2 + 1| + \text{c} = \ln(x^2 + 1) + \text{c}$.

Change of Variables

In the substitution method we picked an expression to be u then checked for du and if it's there we use the power rule or simple recognition to integrate it. With change of variables, which is usually used on radicals, we make the expression inside the radical u then change all other variables to expressions in u including the dx.

Example:

$\int \sqrt{x+1} \cdot x^2 dx$

Let $u = x + 1$ then $du = d(x + 1) = d(x) + d(1) = dx$

Now, since $u = x + 1 \Rightarrow x = u - 1$ and $x^2 = (u - 1)^2 = u^2 - 2u + 1$

So the integral $\int \sqrt{x+1} \cdot x^2\, dx$ becomes $\int \sqrt{u}\, (u^2 - 2u + 1)\, du = \int u^{1/2}\, (u^2 - 2u + 1)\, du$

So each x variable has been replaced with u variables (change of variable method).

$$\int u^{1/2}\, (u^2 - 2u + 1)\, du$$

$$\int (u^{1/2} \cdot u^2 - u^{1/2} \cdot 2u + u^{1/2} \cdot 1)\, du$$

$$\int (u^{5/2} - 2u^{3/2} + u^{1/2}\,)\, du$$

$$\frac{u^{7/2}}{7/2} - \frac{2u^{5/2}}{5/2} + \frac{u^{3/2}}{3/2} + \text{c}$$

$$\frac{2}{7}u^{7/2} - \frac{4}{5}u^{5/2} + \frac{2}{3}u^{3/2} + \text{c}$$

$$\frac{2}{7}(x+1)^{7/2} - \frac{4}{5}(x+1)^{5/2} + \frac{2}{3}(x+1)^{3/2} + \text{c}$$

Integration by Parts

Since the differential of a product is:

$$d(u \cdot v) = udv + vdu$$

We can integrate both sides of the equation and get:

$$\int d(u \cdot v) = \int (udv + vdu)$$

$$u \cdot v = \int u\, dv + \int v\, du$$

$$\int u\, dv = u \cdot v - \int v\, du$$

The idea behind integration by parts is that when given an integral $\int u\, dv$ that our other methods won't solve, we can sometimes integrate it by splitting it up as $u \cdot v - \int v\, du$ where the differential $v\, du$ can be integrated (without a constant c).

Example:

$\int xe^x\, dx$ The x will be u and $e^x dx$ will be dv.

$$u = x \qquad dv = e^x dx$$

$$du = dx \qquad v = e^x$$

So we have,

$$\int xe^x\, dx = u \cdot v - \int v\, du$$

$$= xe^x - \int e^x\, dx$$

$$= xe^x - e^x + \text{c}$$

i.e. $\int xe^x\, dx = xe^x - e^x + \text{c}$

This is a nice method for solving many integrals. The problem is it doesn't always work and when it does, if x in the previous example is x^2 or a higher power, then these same steps would have to be repeated several times. To overcome this problem we use the tabular method. text here.

Tabular Method

To use this method u must be a monomial i.e. an expression whose derivative is eventually zero and of course dv and all its antiderivatives are integrable. We make two columns, one for u in which we continue differentiating until we obtain zero and one for dv in which we continue integrating until even with zero in the u column. We draw diagonal arrows and alternate signs on them as shown. Then combine the linked terms with their sign as the answer.

Example:

$\int x^4 e^x \, dx$

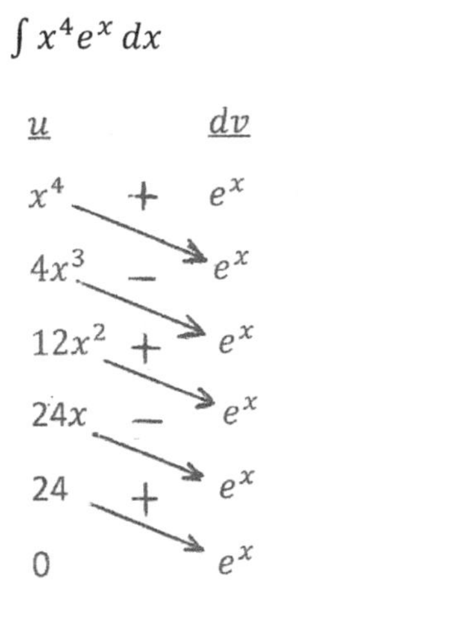

So $\int x^4 e^x \, dx = x^4 e^x - 4x^3 e^x + 12x^2 - 24xe^x + 24e^x$ +c

Summary

As the reader can see these methods seem like just clever manipulations of differentials that facilitate integration. That's because they are. And they have been used for over 300 years. One couldn't "do calculus" unless one learned these "tricks. These are not even all of them. There are many more. At one time scientists and mathematicians even used tables of general form integrals to solve integral problems.

As mentioned previously we now have calculators and computers to do this work for us which frees up time to apply these ideas to actual physical problems without spending hours, days, weeks on calculations

VELOCITY/POSITION

Velocity and position

Previously we saw that velocity is the derivative of the position with respect to time i.e. $v = \frac{dx}{dt}$ so, if we are given a velocity formula we can integrate to a position formula. We can even find the value of c if we are given the velocity formula and initial conditions.

Example:

1) Given that $v(t) = 2t$ and that $x(2) = 11$ find the position formula $x(t)$.

$$\frac{dx}{dt} = 2t$$

$$dx = 2tdt$$

$$\int dx = \int 2tdt$$

$$x + c_1 = t^2 + c_2$$

$$x = t^2 + \text{c} \quad (\text{c} = c_2 - c_1)$$

$$11 = 2^2 + \text{c}$$

$$\text{c} = 7$$

$$\therefore \quad x(t) = t^2 + 7$$

We have also seen that acceleration is the derivative of velocity with respect to time i.e. $a = \frac{dv}{dt}$ So velocity is the antiderivative of acceleration. This means that given an acceleration formula and initial conditions both for velocity as well as position we can find velocity and position formulas.

Example:

2) Given: $a(t) = 24t + 10$, $v(0) = -4$, $x(1) = 6$ find the velocity $v(t)$ and position $x(t)$ formulas.

$$\frac{dv}{dt} = 24t + 10$$

$$dv = (24t + 10)\,dt$$

$$\int dv = \int (24t + 10)\,dt$$

$$v + c_1 = 12t^2 + 10t + c_2$$

$$v = 12t^2 + 10t + c \quad (c = c_2 - c_1)$$

$$-4 = 12(0) + 10(0) + c$$

$$-4 = c$$

$$\therefore \quad v(t) = 12t^2 + 10t - 4$$

$$\frac{dx}{dt} = 12t^2 + 10t - 4$$

$$dx = (12t^2 + 10t - 4)\,dx$$

$$\int dx = \int (12t^2 + 10t - 4)\,dx$$

$$x + c_1 = 4t^3 + 5t^2 - 4t + c_2$$

$$x = 4t^3 + 5t^2 - 4t + c \quad (c = c_2 - c_1)$$

$$6 = 4 + 5 - 4 + c$$

$$1 = c$$

$$\therefore \quad x(t) = 4t^3 + 5t^2 - 4t + 1$$

AREA

Differential Properties

We have been using the infinitesimal differentials dx, dy, ds to find related rates, tangents, velocity and optimization. We said previously that differentials are smaller than any positive real number so therefore they neither increase nor decrease the value of real numbers when added to or subtracted from them i.e.

$$r \mp dx = r$$

Higher power differentials like dx^2 are infinitely smaller than dx and are effectively zero, so they can be dropped from any expression containing them

This property will be used in our discussion of the ultimate parts of area dA.

Area Differentials

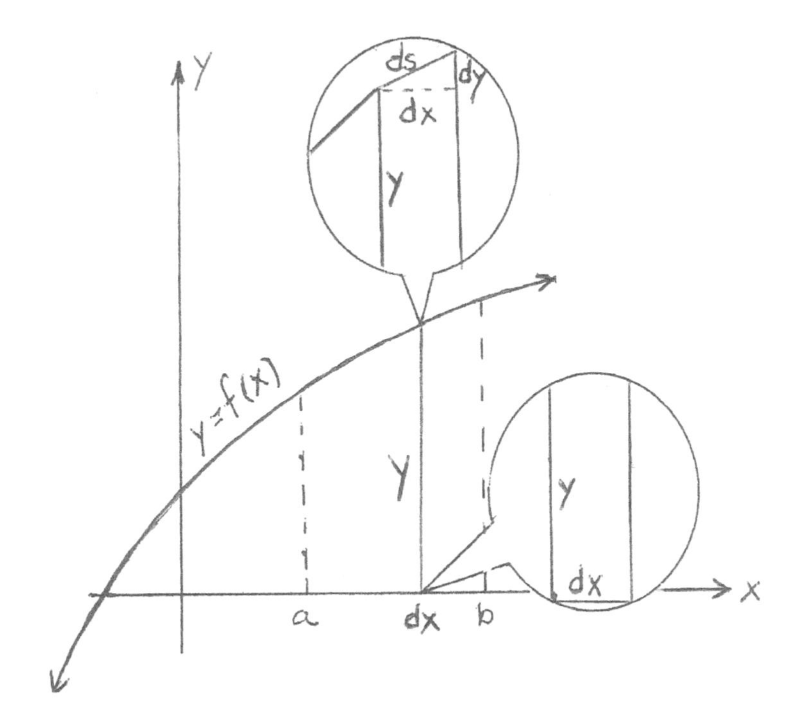

If we wish to find the total area under a curve from $x = a$ to $x = b$, we can imagine infinitely many infinitely narrow trapezoids with their bases on the x-axis. Each having width dx and length y up to the curve. The top triangular part has hypotenuse ds, the lineal portion of the curve, and sides dx and dy. The rectangular portion has length y and width dx. The area of each rectangle is ydx which is infinitesimal.

Note that the infinitesimal triangle (the characteristic triangle) atop the rectangle ydx has hypotenuse ds of the curve and area $\frac{1}{2} \cdot dx \cdot dy$. This product is of the same order as dx^2 so it is effectively zero and may be omitted. This leaves only the rectangles each with area ydx to consider. We can say then that the ultimate parts of area dA under the curve are just these infinitely narrow rectangles and we use the integral to "sum up" all of these infinitesimal parts of area dA to give us the total area A.

$$dA = ydx$$

$$\int dA = \int ydx$$

$$A = \int ydx$$

This is the reason Leibnitz chose the symbol $\int$ it looks like an elongated s for sum, it's the sum of the differentials of area. But we also know that $\int$ is also an anti-differential so to "sum up" the dA's we find the anti-differential of dA which is the anti-differential of ydx. We will call this $Y(x)$ + c

So, when finding an area below a curve from an x-value a to another x-value b we use a "definite integral" which we write as $\int_a^b y\,dx = Y(b) - Y(a)$ where Y is the anti-differential of ydx i.e $dY = ydx$. We also use $Y|_a^b$ to symbolize $Y(b) - Y(a)$. Note that the constant c subtracts out in the process because $(Y(b) + c) - (Y(a) + c)$ leaving only $Y(b) - Y(a)$.

Example: Find the area under the curve $y = \sin x$ from $x = 0$ to $x = \pi$

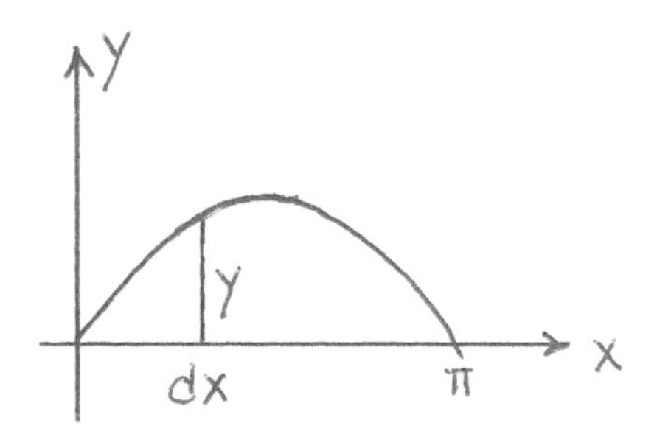

$dA = ydx$ and since $y = \sin x$, $dA = \sin x\,dx$. So the area from $x = 0$ to $x = \pi$ is:

$$A|_0^\pi = \int_0^\pi \sin x\,dx = -\cos x\,|_0^\pi$$

$$= -(\cos\pi - \cos 0) = 2$$

Example: Find the area under the curve $y = e^x$ and above the x - axis from $x = 0$ to $x = 1$

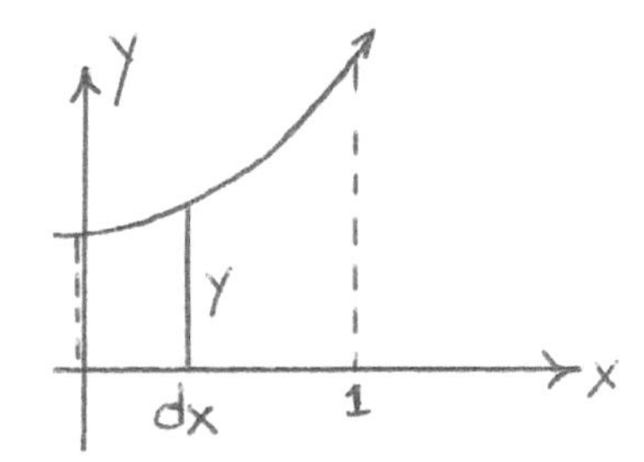

$dA = ydx = e^x dx$ since $y = e^x$ So, the area from $x = 0$ to $x = 1$ is:

$$A|_0^1 = \int_0^1 e^x\,dx = e^x\,|_0^1 = e^1 - e^0 = e - 1$$

In standard calculus the definite integral is written as $\int_a^b f(x)\,dx$. The a and b are called the "limits of integration," $f(x)$ is called the integrand and $F(b) - F(a)$ is how it's evaluated. $F(x)$ is called the anti-derivative of $f(x)$. The limits of integration are x values when the differential is dx, they are y values when the differential is dy. This equation, $\int_a^b f(x)\;dx = F|_a^b = F(b) - F(a)$ is called the Fundamental Theorem of Calculus. It was discovered by both Newton and Leibnitz independently.

Also, in standard calculus the meaning of $\int_a^b f(x)\,dx$ is defined as the limit of a Riemann sum i.e.

$$\int_a^b f(x)\,dx = \lim_{n\to\infty} \sum_{i=1}^{n} f(x_i)\,\Delta x_i$$

The Riemann sum is applied to the rectangles of height $f(x_i)$ and width Δx_i of a partition. In fact most calculus texts devote several pages to a discussion of the various types of partitions; left justified, right justified, and midpoint, then equal and unequal widths of partitions. Often the "norm" of the partition is also discussed as well as the relationship between the limit of the norm of the partition approaching zero and the number of rectangles approaching infinity. The student is then asked to set up partitions and use the limit of their Riemann sums to find area. After a sufficient amount of torture, the student is then shown how this limit ties into the Fundamental theorem of Calculus and a collective sigh of relief is heard when they realize that area can be found with integrals instead. All of this is beautiful mathematics to the mathematician but trauma to most students.

Properties of Definite Integrals

1) $\int_a^b (f(x) \mp g(x))\;dx = \int_a^b f(x)\;dx \mp \int_a^b g(x)\,dx$

2) $\int_a^b kf(x)\,dx = k\int_a^b f(x)\,dx$

3) $\int_a^a f(x)\,dx = 0$

4) $\int_a^b f(x)\,dx = -\int_b^a f(x)\;dx$

5) If $a \le b \le c$ then $\int_a^c f(x)\,dx = \int_a^b f(x)\,dx + \int_b^c f(x)\,dx$

Horizontal Strips

We have been using vertical $dA's$. Sometimes it may be easier to use horizontal dA's. In this case we will have rectangles of length x and width dy. So $dA = xdy$ and $A = \int xdy = X(y) + \text{c}$.

Example: Find the area of the region above the curve $y = x^2$ and below the horizontal line $y = 4$ between $y = 0$ and $y = 4$.

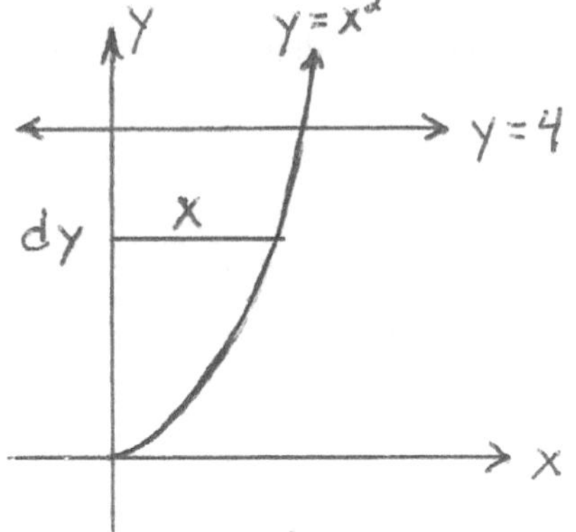

$dA = xdy$. Since the width of each dA is dy we must write x in terms of y. Since $y = x^2$, $x = \sqrt{y}$ so $dA = \sqrt{y}\,dy$ (recall that the variable must match the differential). And the area is:

$$A|_0^4 = \int_0^4 \sqrt{y}\,dy = \frac{2}{3}y^{3/2}|_0^4 = \frac{16}{3}.$$

Sometimes changing the variable to match the differential may be very difficult or impossible. If that's the case it may be easier to change the differential to match the variable. This is called a "formal approach" because now our formula would not correspond to our picture.

Solving the example above in this fashion we first find dy. Since $y = x^2$, $dy = 2xdx$ and for $dA = xdy$ we substitute and get $dA = x(2xdx) = 2x^2dx$. Since we have differential dx we change the y values 0 and 4 to the x values $x = 0$ and $x = 2$ using $x = \sqrt{y}$. We then have:

$$A|_0^2 = \int_0^2 2x^2dx = \frac{2}{3}x^3|_0^2 = \frac{16}{3}$$

This is the same answer we got with horizontal dA's. But in this case $dA = 2x^2dx$ is not a horizontal or vertical dA in the region we want. It is just a formal expression obtained by substitution that is used to find the desired area. Once we are empowered to think of differentials as meaningful entities in and of themselves we open up new vistas for using them to solve problems.

This same problem can also be solved with vertical dA's. The length of each such dA would be $(4 - y)$ and the width would be dx. We substitute x^2 for y and $dA = (4 - x^2)dx$ and the area is:

$$A|_0^2 = \int_0^2 (4 - x^2)dx = 4x - \frac{x^3}{3}|_0^2 = \frac{16}{3}$$

Example: (vertical strips) Find the area between $x = 1$ and $x = 3$ under $y = x^3$ and above the $x - axis$.

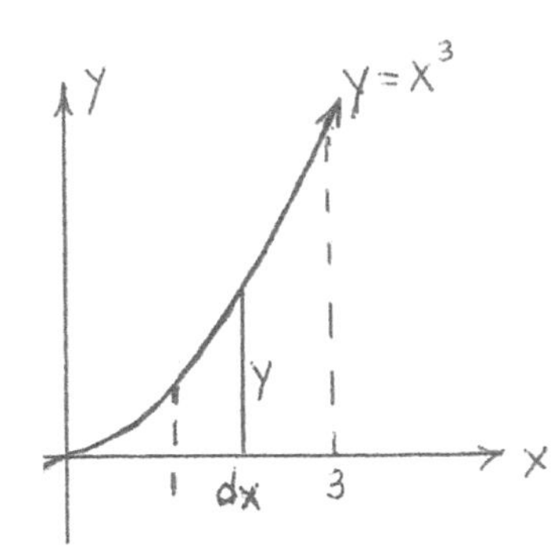

$dA = ydx$

$A|_1^3 = \int_1^3 ydx = \int_1^3 x^3dx$

$= \frac{x^4}{4}|_1^3$

$= \frac{81}{4} - \frac{1}{4}$

= 20

Example 2): (horizontal strips) Find the area between $y = 0$ and $y = 4$ above $y = \sqrt{x}$ below the line $y = 4$

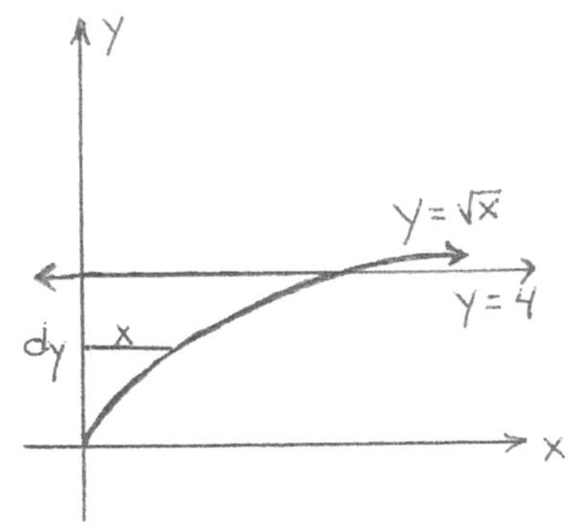

$dA = xdy$

Since $y = \sqrt{x} \Rightarrow x = y^2$

So $dA = y^2dy$

$A|_0^4 = \int_0^4 y^2dy$

$= \frac{y^3}{3}|_0^4$

$= \frac{64}{3}$

Example 3) Find the area between $y = 0$ and $y = ln\,4$ and between the $y - axis$ and $y = ln\,x$. Horizontal strips first.

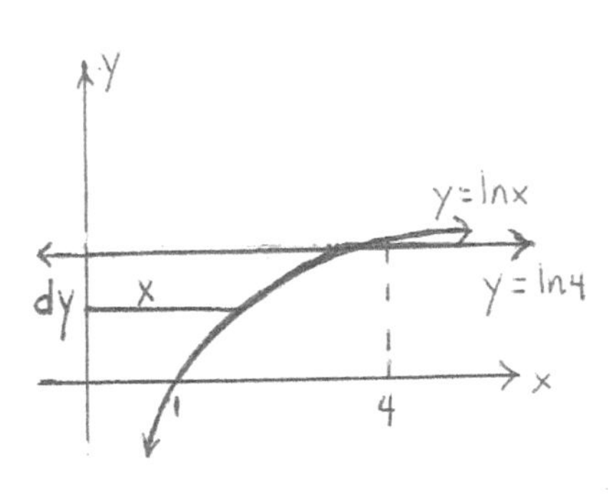

$dA = xdy$

Since $y = ln\,x \Rightarrow x = e^y$

So $dA = e^ydy$ and

$A|_0^{ln4} = \int_0^{ln4} e^ydy = e^y|_0^{ln4}$

$= e^{ln4} - e^0$

= 4 - 1

= 3

Same example 3 but change differentials instead:

$$dA = xdy \quad \text{now since } y = \ln x \Rightarrow dy = \frac{dx}{x}$$

So $dA = x\frac{dx}{x} \Rightarrow dA = dx$ also, since $y = 0$ and $y = \ln 4 \Rightarrow x = 1$ and $x = 4$

we change from y to x limits of integration $A|_0^{\ln 4} = A|_1^4 = \int_1^4 dx$

$= x|_1^4$

$= 4 - 1$

$= 3$

Negative Regions

If a region is below the x-axis then the differential of area $dA = ydx$ will be negative since y represents negative values. So to make it positive we prefix it with a negative sign. So $-ydx$ will be positive and dA will be positive.

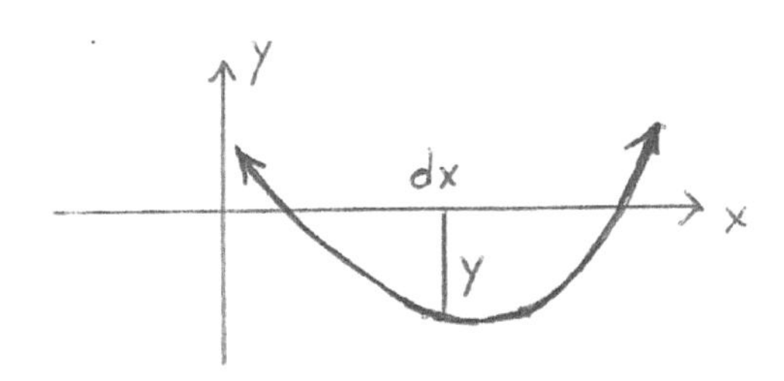

Likewise when a region is to the left of the y - axis $dA = xdy$ would be negative, since x represents negative values. So $-xdy$ will be positive and dA will again be positive.

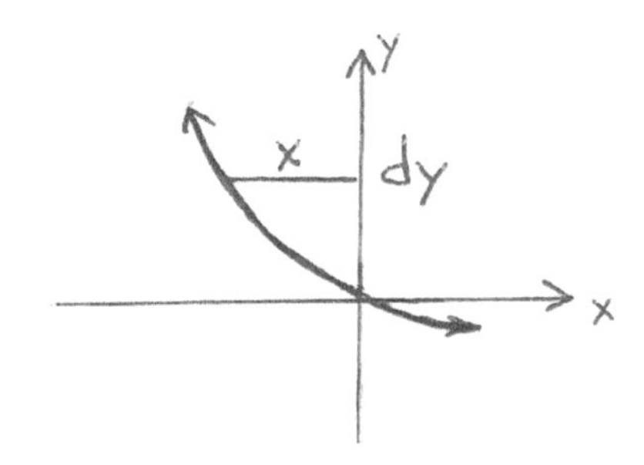

Area Between Two curves

When a region is between the curve and *x-axis* (or *y-axis*) then the length of dA is just y (or x). To find the area between two curves then y = $y_2 - y_1$ (or $x_2 - x_1$). That is, the upper minus the lower curve or the right minus the left curve.

Examples:

1) Find the area between $y_2 = x$ and $y_1 = x^2$ from $x = 0$ to $x = 1$

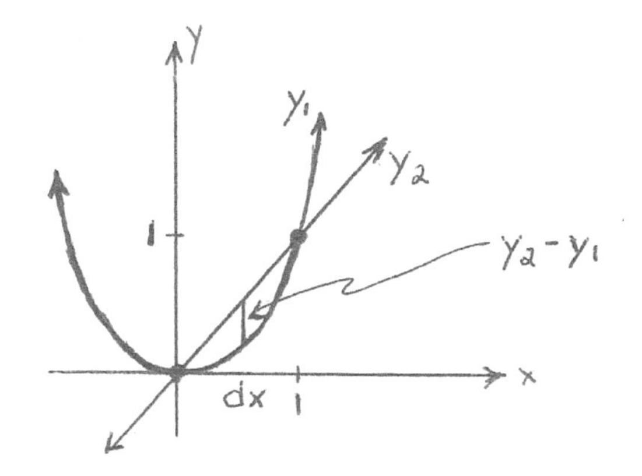

$dA = (y_2 - y_1)dx = (x - x^2)dx$

So $A|_{x=0}^{x=1} = \int_0^1 (x - x^2)dx$

$= \frac{x^2}{2} - \frac{x^3}{3} |_0^1$

$= \frac{1}{6}$

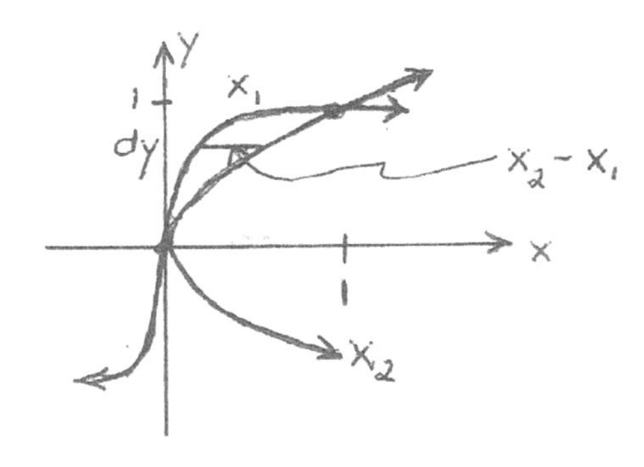

2) Find the area between $x_2 = y^2$ & $x_1 = y^3$ from y = 0 to y = 1

$dA = (x_2 - x_1)dy = (y^2 - y^3)dy$

So $A|_{y=0}^{y=1} = \int_0^1 (y^2 - y^3)dy$

$= \frac{x^3}{3} - \frac{x^4}{4} |_0^1 = \frac{1}{12}$

The Rate of Area Change

Calculating an area that would have taken the genius of an Archimedes to find in the ancient world can now be found by high school students with a rudimentary knowledge of integral calculus. Now the problem is made even simpler because many cheap scientific calculators can solve definite integrals. Some calculators even provide formulaic differential and integral solutions! So once a student is able to represent a problem as a definite integral, the solution is even quicker.

Interesting note regarding integrals. Recall that an "ultimate part" or differential of area is:

$$dA = \text{y}\, dx \quad \text{now divide by } dx$$

$$\frac{dA}{dx} = y$$

This expression tells us that the rate the area changes with respect to x at a location of the curve is the y value of the location.

Net Change

We have seen that an integral can be understood as the infinite sum of the ultimate parts of area that gives us the total area. We have also seen that care must be taken to determine whether or not the differentials were negative so that we can turn them positive when finding area. But if we integrate $\text{y}'dx$ from $x = a$ to $x = b$ without concern for sign, we are instead finding the net change in y values i.e $\int_a^b \text{y}'dx = \int_a^b \frac{dy}{dx} dx = \int_a^b dy = \text{y(b)} - \text{y(a)}$ is the net change in y from $x = a$ to $x = b$.

So integrals can be used to find area and net change in variables. They can also be used to find volume which is our next topic.

VOLUME

Disc Method

If we revolve a planar region about an axis we create a "solid of revolution." We can think of its volume as consisting of infinitely many infinitely thin "discs" of volume dV. These are the ultimate parts of volume which are then "summed up" with an integral to get the total volume. A disc of volume dV is really a flattened cylinder. Since the volume of a cylinder is $V = \pi r^2 h$, the volume of a flattened cylinder disc is $dV = \pi r^2 dh$.

Example: Find the volume of the solid formed by revolving the region above y = x^2 and to the right of the y – axis, from $y = 0$ to $y = 4$ about the y – axis.

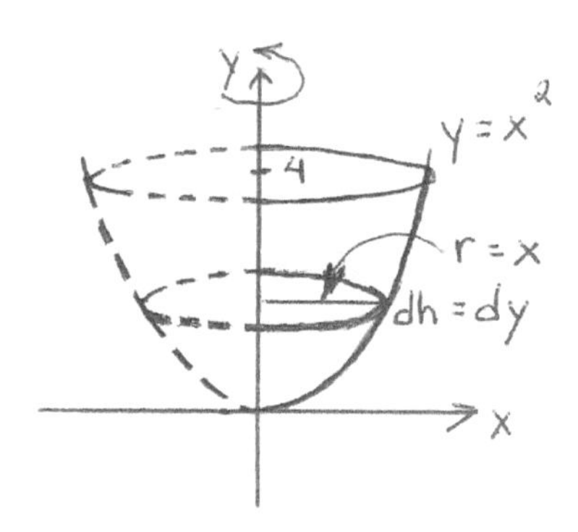

In this case $r = x$ and $dh = dy$. So:

$dV = \pi x^2 dy$. Now substitute y for x^2 to match dy and $dV = \pi y dy$

And then the total volume is $V|_0^4 = \int_0^4 \pi y dy = \frac{\pi y^2}{2}|_0^4 = \frac{16\pi}{2} = 8\pi$

Example: Find the volume of the solid formed by revolving the region below y = x^2 and above x-axis from $x = 0$ to $x = 2$ about the x-axis.

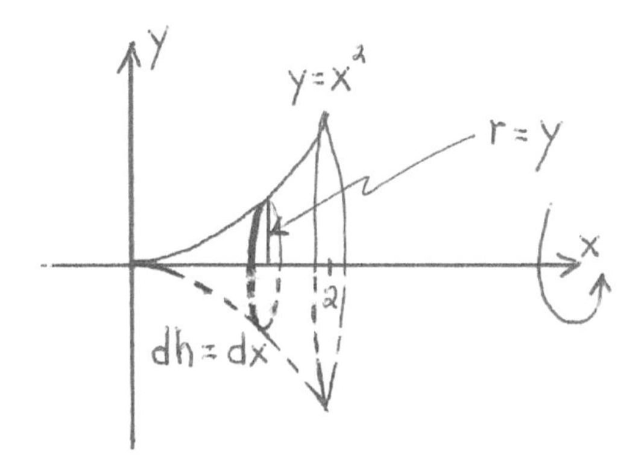

In this case $r = y$ and $dh = dx$. So:

$dV = \pi y^2 dx$. Now substitute x^4 for y^2 to match dx and we get $dV = \pi x^4 dx$.

And the volume is $V|_0^2 = \int_0^2 \pi x^4 dx = \pi \frac{x^5}{5}|_0^2 = \frac{32\pi}{5}$

The Washer Method

If a region is between two curves and is revolved about an axis it will usually create a volume with a hole in it. In this case, instead of discs we use washers.

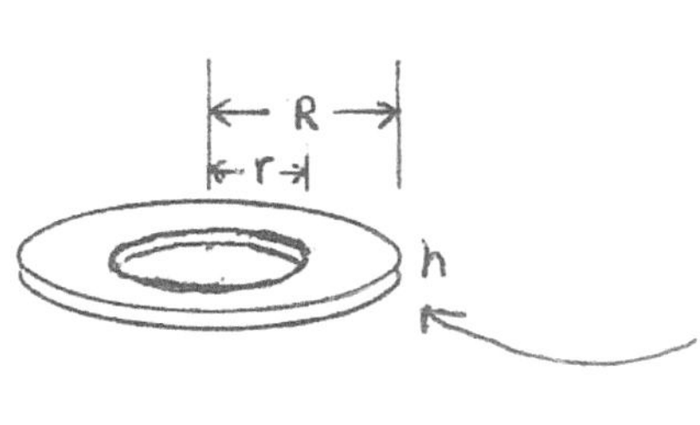

$V_{disc} = \pi R^2 h$

$V_{hole} = \pi r^2 h$

So the volume of the washer would be

$V_{washer} = \pi R^2 h - \pi r^2 h = \pi(R^2 - r^2)h$

A washer differential would be

$dV = \pi(R^2 - r^2)dh$

Example: Find the volume of the solid generated by revolving the region between y = x_1^2 and y = x_2 about the $y - axis$

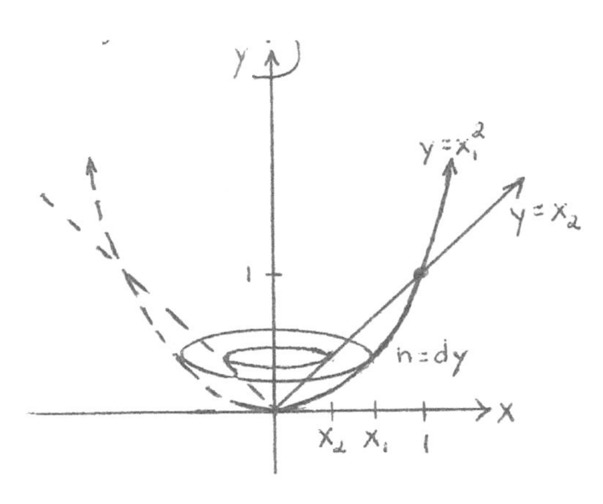

$R = x_1$ $r = x_2$ $dh = dy$ $x_1 = \sqrt{y}$ and x_2 = y

In this problem the y values are equal and, $x_1 = \sqrt{y}$ and $x_2 = y$

So, $dV = \pi(x_1{}^2 - x_2{}^2)dy = \pi(\sqrt{y}^2 - y^2)dy = \pi(y - y^2)dy$ And

$V|_0^1 = \pi \int_0^1 (y - y^2)dy = \pi(\frac{y^2}{2} - \frac{y^3}{3})|_0^1$

$= \frac{\pi}{6}$

The Shell Method

Solids with holes (and even without) can also be evaluated using the shell method. A shell or "can" is nothing more than a rectangular sheet with length = $2\pi r$, height = h, and thickness = w that is wrapped around to form a shell as shown below. The volume of this rectangular shell sheet is $V = 2\pi rhw$. The volume of the solid that is formed with a rotation is composed of infinitely many infinitely thin shells placed one within another until the shells form a solid with a hole in the middle. The height and radius will vary from shell to shell but the width of each shell will always be infinitesimal i.e. w will be dw. This dw will be either dx or dy (perpendicular to the axis of rotation.) Each infinitely "thin" shell has volume $dV = 2\pi rhdw$.

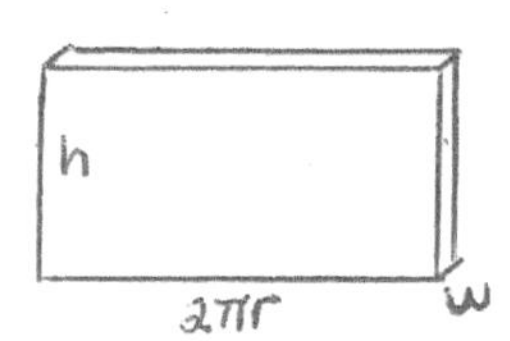

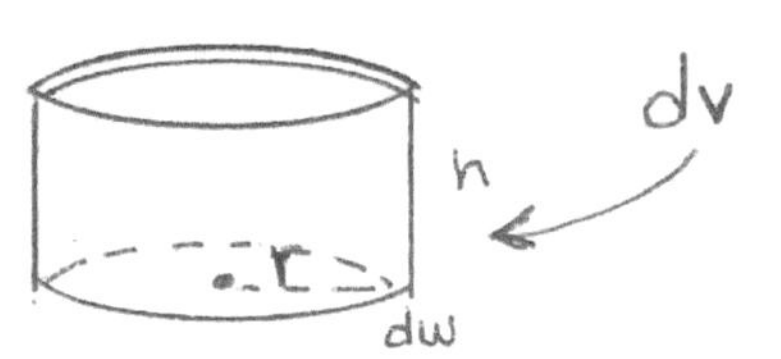

Example: Find the volume of the solid formed by revolving the region enclosed by $y = \sin x$ and the x - *axis* from $x = 0$ to $x = \frac{\pi}{2}$ about the y - *axis.*

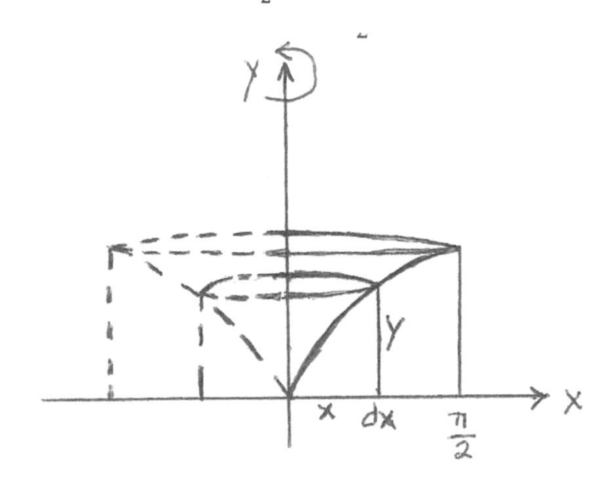

$r = x \quad h = y \quad dw = dx$ So $dV = 2\pi xydx$

Substitute $y = \sin x$ and $dV = 2\pi x \sin x \, dx$

So $V|_0^{\pi/2} = 2\pi\int_0^{\pi/2} x \sin x \, dx = 2\pi$

Rotation About Other Lines

A curve may be rotated about another line besides the axes. If it is a vertical line l then if using discs the radius would be $r = l - x$ or $r = l - y$ depending on whether it is a horizontal or vertical line respectively as shown below. And $dh = dx$ or dy So, generally, $dV = \pi(l - x)^2 dh$

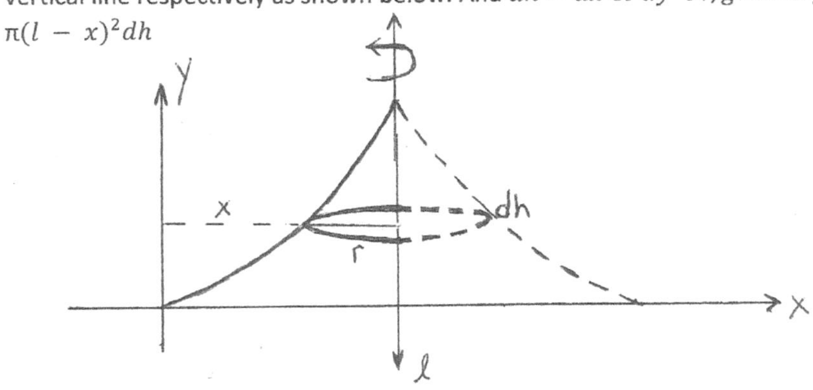

Example: Find the volume of the solid generated when revolving the region between the $x-$ *axis* the, curve $y = x,^2$ and the line $x = 1$ about the line $x = 1$

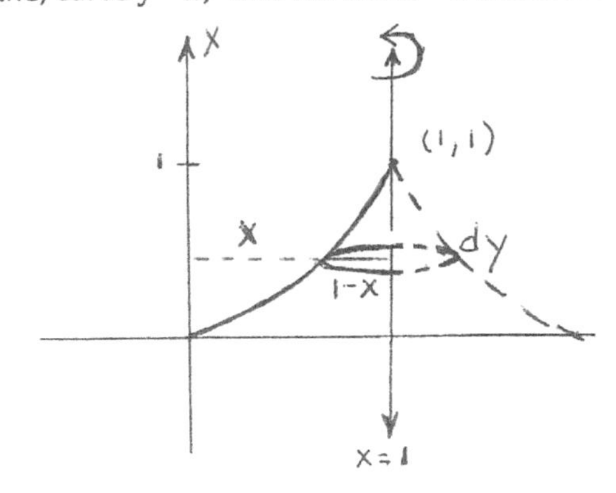

$r = 1 - x$ $dh = dy$ $x = \sqrt{y}$ and

$dV = \pi(1 - x)^2 dy =$

$\pi(1 - \sqrt{y})^2 dy$ So the volume from $y = 0$ to $y = 1$ is

$V|_0^1 = \int_0^1 \pi(1 - \sqrt{y})^2 dy$

$= \frac{\pi}{6}$

Volumes by Cross-Sections

Sometimes we are asked to find the volume of a solid that is not the result of a "revolution." The base may be an ellipse, circle, parabola or any basic function. And the solid itself may have cross-sections that are always triangles, or rectangles with varying sizes that vary with the base and cross-section. But their thickness is infinitesimal. The total volume will be the sum of these "ultimate parts" of volume. The volume of each cross-section will be the area, which will be symbolized as $A(x)$ or $A(y)$, opposite of the axis it is perpendicular to, multiplied by the thickness dw which will be dx or dy, the same as the axis the cross-section is perpendicular to. We will then have either $dV = A(x)dy$ or $dV = A(y)dx$.

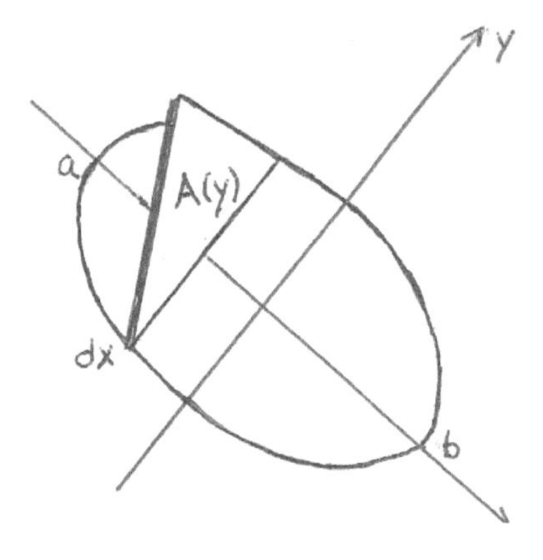

In this diagram have:

Triangular cross-sections with area $A(y)$ and thickness $dw = dx$

So $dV = A(y)dx$ stacked from $x = a$ to $x = b$. And the total volume will be,

$V|_a^b = \int_a^b A(y)dx$ Of course $A(y)$ will be replaced with an area in terms of y then re-expressed in terms of x since we have dx.

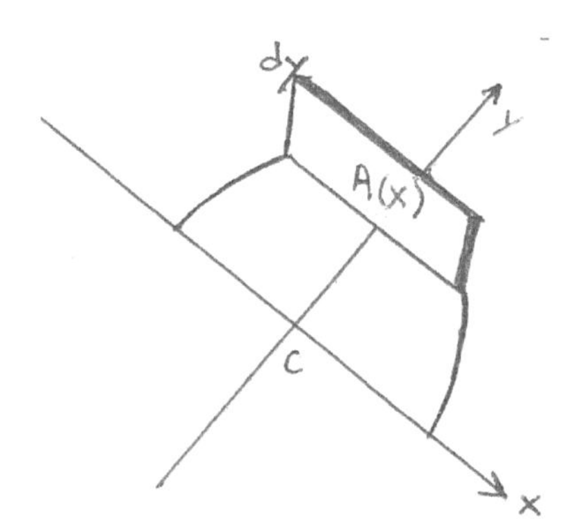

In this diagram we will have:

Rectangular cross-sections with area $A(x)$ and thickness $dw = dy$

So $dV = A(x)dy$ stacked from $y = c$ to $y = d$. And the total volume will be,

$V|_c^d = \int_c^d A(x)dy$ Of course $A(x)$ will be replaced with an area in terms of x then re-expressed in terms of y since we have dy.

Example 1) Find The volume of the solid with base $y = \cos x$ from $x = \frac{-\pi}{2}$ to $x = \frac{\pi}{2}$ with isosceles right triangle cross-sections and leg perpendicular to the $x - axis$.

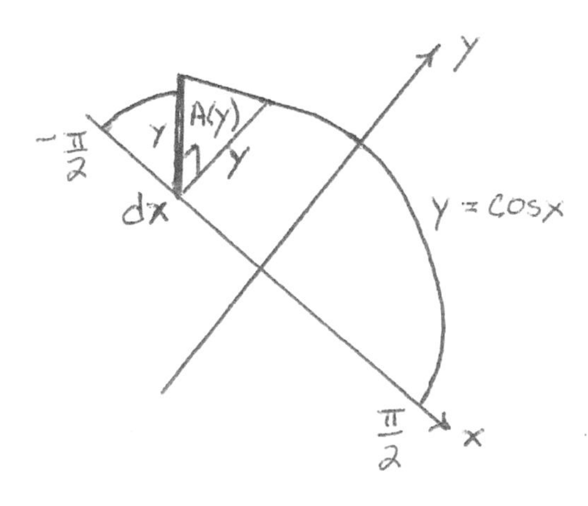

The area of the isosceles right triangles is a function of y so we have:

$A(y) = \frac{1}{2} y \cdot y = \frac{y^2}{2}$ and $dw = dx$ so $dV = \frac{y^2}{2} dx = \frac{(\cos x)^2}{2} dx$

$V|_{-\pi/2}^{\pi/2} = \frac{1}{2}\int_{-\pi/2}^{\pi/2} (\cos x)^2 \, dx$

$= \frac{\pi}{4}$

Example 2) Find The volume of the solid with base $y = x^2$ from $y = 0$ to $y = 4$ with semi-circle cross-sections perpendicular to the $y - axis$.

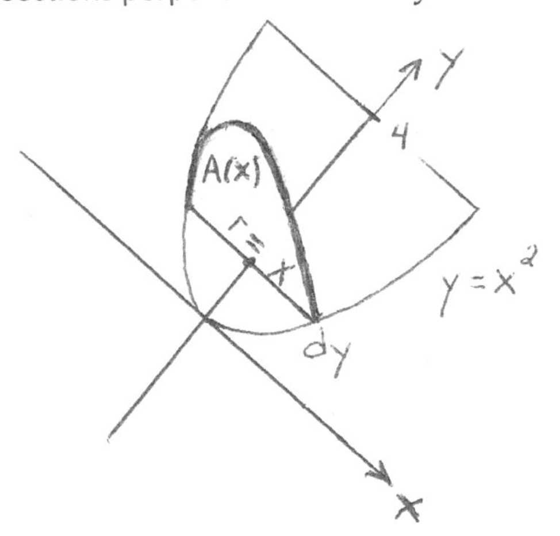

The area of the semi-circle is a function of x so we have:

$A(x) = \frac{1}{2} \pi x^2$ and $dw = dy$ so $dV = \frac{1}{2} \pi y \, dy$

$V|_0^4 = \frac{\pi}{2} \int_0^4 y \, dy$

$= 4\pi$

Arc Length

We have seen that we can use differentials to represent the "ultimate parts" of area and volume, then use an integral to sum up all of these parts to get the total area or volume. We can do the same thing to find arc length. We have already discussed the fact that the ultimate parts of a curve are actually infinitesimal lineal segments ds that we called micro-segments. To find the arc length s then, all we have to do is sum up all of the ds. So we need a way to represent ds in terms of dx and dy. From Leibnitz characteristic right triangle we see that:

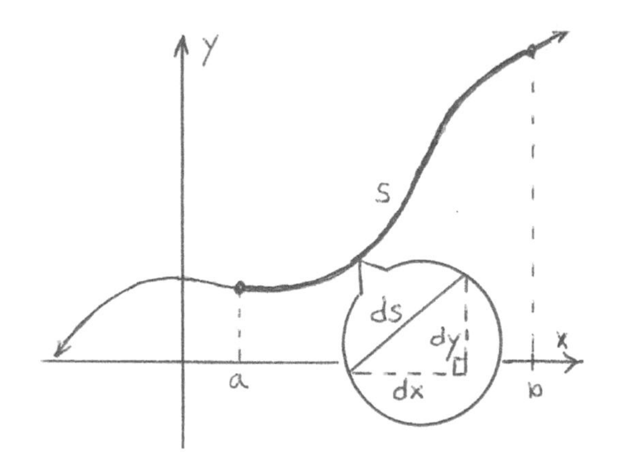

$$ds^2 = dx^2 + dy^2$$

$$\frac{ds^2}{dx^2} = \frac{dx^2}{dx^2} + \frac{dy^2}{dx^2}$$

$$\frac{ds}{dx} = \sqrt{1 + (y')^2}$$

$$ds = \sqrt{1 + (y')^2}\ dx$$

$$\int ds = \int \sqrt{1 + (y')^2}\ dx$$

$$s = \int \sqrt{1 + (y')^2}\ dx$$

$$s|_a^b = \int_a^b \sqrt{1 + (y')^2}\ dx$$

$s|_a^b$ represents the arc length from $x = a$ to $x = b$

Example: Find the arc length from $x = 0$ to $x = 1$ of the curve y = x^2

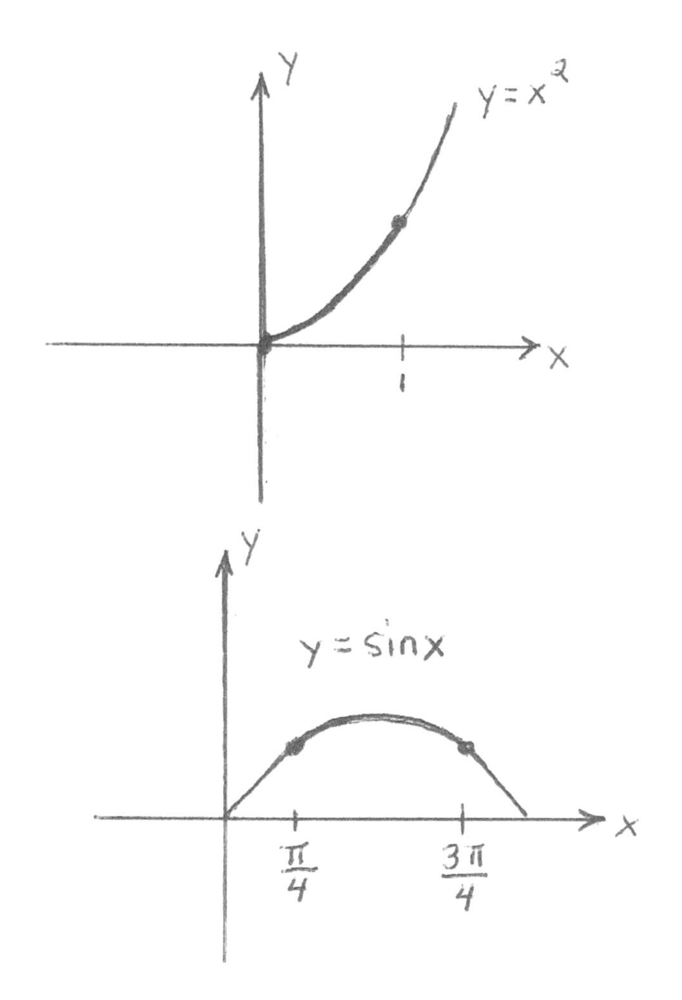

$y' = \frac{dy}{dx} = \frac{d(x^2)}{dx} = \frac{2xdx}{dx} = 2x$

$s|_0^1 = \int_0^1 \sqrt{1 + (2x)^2}\, dx$

≈ 1.48

Example: Find the arc length from $x = \frac{\pi}{4}$ to $x = \frac{3\pi}{4}$ of the curve y = $\sin x$

$y' = \frac{dy}{dx} = \frac{d(\sin x)}{dx} = \frac{\cos dx}{dx} = \cos x$

$s|_{\pi/4}^{3\pi/4} = \int_{\pi/4}^{3\pi/4} \sqrt{1 + (\cos x)^2}\, dx$

≈ 1.7

Surface Area

To find surface area we need to review a frustum. In geometry we learn that the surface area S of a frustum is:

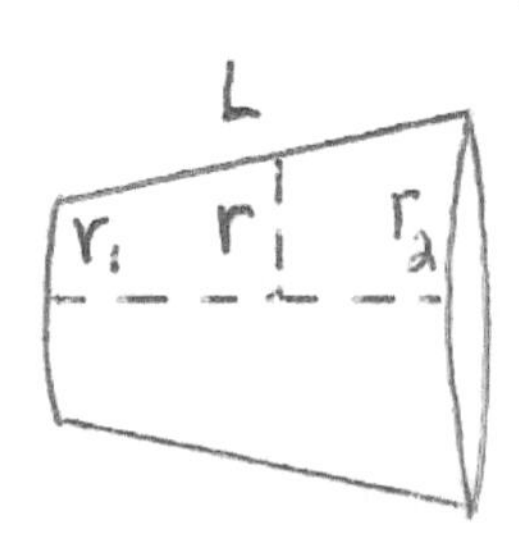

$$S = \pi\,(r_1 + r_2)\,L$$

$$= 2\,\pi\,(\frac{r_1 + r_2}{2})\,L$$

$$= 2\,\pi\,r\,L$$

If we want to find the surface area of a solid and we use only one frustum it would be very inaccurate. But if we sum up the ultimate parts of surface area i.e. frustums with $L = ds$ then we won't have any error. A frustum with $L = ds$ and $r = y$ will be a differential of surface area $dS = 2\pi y\ ds$. We know from our unit on arc length that $ds = \sqrt{1 + (y')^2}\ dx$ so:

$$dS = 2\pi y\ \sqrt{1 + (y')^2}\ dx$$

$$S|_a^b = \int_a^b 2\pi y\sqrt{1 + (y')^2}dx$$

Example: Find the surface area of the solid from $x = 1$ to $x = 3$ formed by revolving $y = x^2$ about the x-*axis*

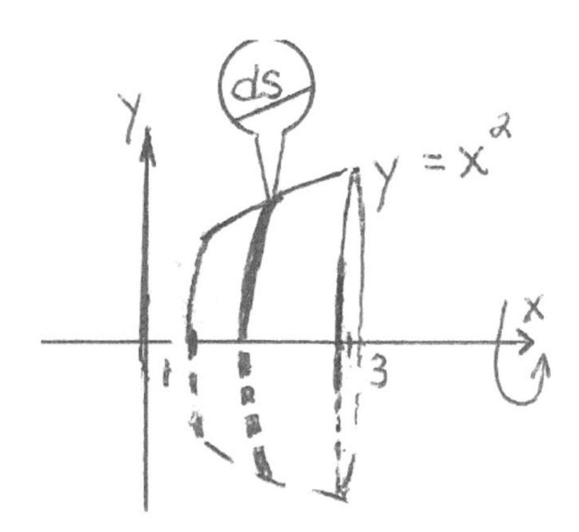

$$S|_1^3 = \int_1^3 2\pi y\sqrt{1 + (y')^2}\ dx$$

$$S|_1^3 = \int_1^3 2\pi x^2\sqrt{1 + (2x)^2}\ dx$$

$$= 257.5$$

MOTION

Velocity

Before discussing horizontal and vertical motion we will compare average and instantaneous velocity. In elementary science class we learn that velocity is a vector. This means that it can be represented with an "arrow." This arrow has both magnitude (speed) and direction.

The average velocity of an object in motion is the distance traveled divided by the time elapsed. Symbolically:

$$v_{avg} = v\big|_{t_1}^{t_2} = \frac{\text{finite distance traveled}}{\text{finite time elapsed}} = \frac{x(t_2) - x(t_1)}{t_2 - t_1} = \frac{\Delta x}{\Delta t}\big|_{t_1}^{t_2}$$

Instantaneous velocity is the infinitesimal distance traveled divided by the instant of time and t_1 represents the only real number valued time in that instant. Symbolically:

$$v(t_1) = \frac{\text{infinitesimal distance traveled}}{\text{one instant}} = \frac{dx}{dt}\big|_{t=t_1}$$

Example: If a particle is moving on the *x-axis* so that its position x is given by $x = t^3 + t^2$ (measured in feet and seconds) its v_{avg} during the first five seconds is:

$$v\big|_0^5 = \frac{x(5) - x(0)}{5-0} = \frac{150-0}{5} = 30\ \frac{ft}{sec}$$

If the particle hits a barrier when $t = 5$, then what is its impact velocity? Now we need the instantaneous velocity for $t = 5$ i.e. $v(5)$.

$$v(5) = \frac{dx}{dt}\big|_{t=5} = \frac{d(t^3 + t^2)}{dt}\big|_{t=5} = \frac{3t^2dt + 2tdt}{dt}\big|_{t=5} = 3t^2 + 2t\ \big|_{t=5} = 3(5)^2 + 2(5) = 85\ \frac{ft}{sec}$$

Clearly the impact velocity is not the same as the average velocity! In fact the particle will have a different velocity at every instant of its motion. The only time period during which the velocity is uniform is during an instant. During an instant it has no accelerations or decelerations, a concept widely used in classical mechanics called micro-uniformity.

Note the difference even in the symbolism between average and instantaneous velocity. Average velocity is between two different instants, t_2 and t_1 whereas instantaneous velocity is at one instant t_1.

Horizontal Motion

If a particle is moving on the x-axis then its instantaneous velocity is $v = \dot{x}$ Velocity is a vector so it has both magnitude (speed) and direction (positive is right, negative is left). If $v = 0$ then the particle is motionless. With this information we can determine the speed and direction of the particle from $\dot{x}$

Example: The position of a particle moving on the x-axis is given by $x = \frac{t^3}{3} - 4t^2 + 12t + 3$. Find the velocity equation and determine whether it's moving right or left.

In order to determine the requested information we find $\dot{x}$ and draw a time line. We plot the critical points on the timeline then determine the sign of v between the critical values (using $\dot{x}$ for any value between the critical values). Depending on the sign, we indicate the direction during those times.

$$v = \dot{x} = \frac{dx}{dt} = \frac{d\left(\frac{t^3}{3} - 4t^2 + 12t + 3\right)}{dt} = \frac{t^2dt - 8tdt + 12dt}{dt} = t^2 - 8t + 12 = (t-2)(t-6) = 0 \text{ when } t = 2 \text{ or } t = 6$$

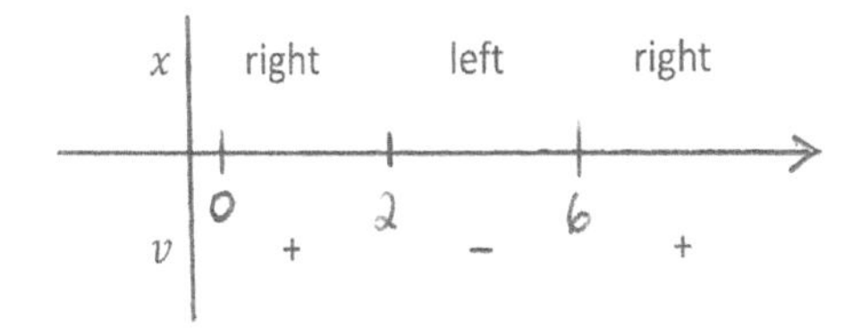

The particle stops when $t = 2$ and $t = 6$ (to change directions).

Acceleration/Deceleration

We have stated that during an instant an object will travel with uniform velocity. It does not accelerate or decelerate. We called this micro-uniformity. But since acceleration/deceleration is $a = \frac{dv}{dt}$ it is clear that during the instant dt the velocity v does change by the amount dv. But this does not change the real-value of v during that instant because dv is infinitesimal and $v + dv = v$. However, as these $dv's$ accumulate over time, they will increase or decrease the real-valued velocity.

Now, acceleration/deceleration a is also a vector. If a works in the same direction as v the object accelerates. If it's in the opposite direction as v the object decelerates. If $a = 0$, the velocity is constant. Recall that the acceleration can also be represented as $\ddot{x}$

So to summarize, if the signs of v and a are the same, then the object is accelerating, if they are different then it's decelerating. The acceleration/deceleration formula alone does not provide enough information to determine this, we need to know the sign of both v and a. We find the critical values of both, plot them on a timeline, and check the signs of these derivatives at any values between them.

Example: The position of a moving particle is given by $x = \frac{t^3}{3} - 4t^2 + 12t + 3$ (this is the same equation as on the previous page where we already found v)

$v = t^2 - 8t + 12 = (t-2)(t-6) = 0$ when $t = 2$ or $t = 6$

$a = \dot{v} = \frac{dv}{dt} = \frac{d(t^2 - 8t + 12)}{dt} = \frac{2tdt - 8dt}{dt} = 2t - 8 = 2(t-4) = 0$ when $t = 4$

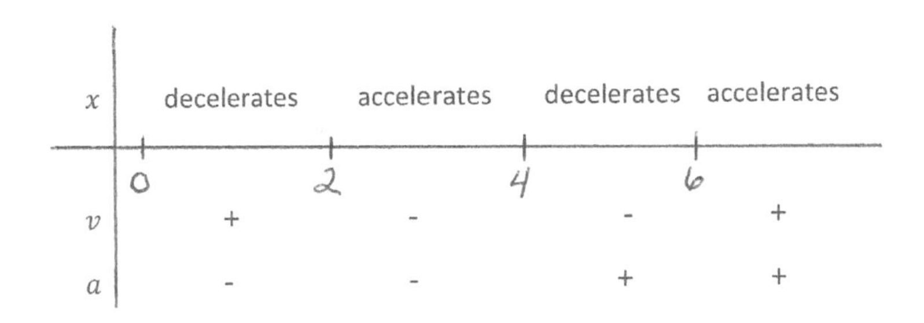

Distance and Integrals

Recall that an integral can represent the area below a curve (and its negative if the region is below the horizontal axis). Position (distance) is the integral of velocity, that means that the area below a velocity curve must be the distance traveled. So if we integrate a velocity formula from $t = a$ to $t = b$ (making negative dA's positive) that will give us the total distance traveled between those time values.

Example: The velocity of a particle moving on the x-axis is given by $v(t) = t^2 - 5t + 6$ find the distance traveled from $t = 0$ to $t = 3$. The graph of v against time t is below.

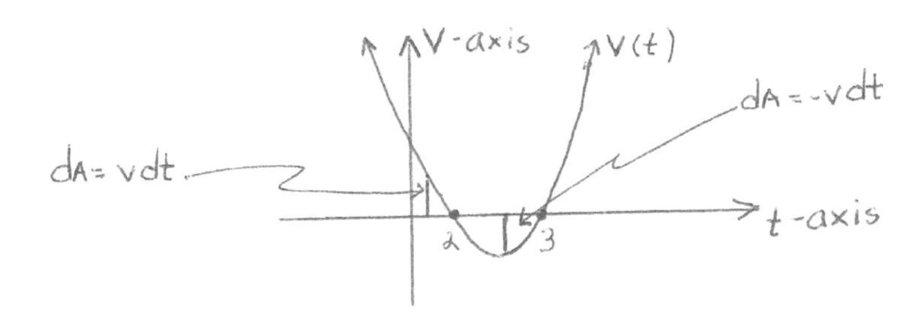

We can do this without drawing the velocity graph as above. Instead we can determine where the velocity is positive or negative on a time axis only (like we did in the previous example) then just form our integrals with the right sign as shown below.

$v(t) = t^2 - 5t + 6 = (t - 3)(t - 2) = 0$ when t = 2 or 3 Check the sign of v between these values

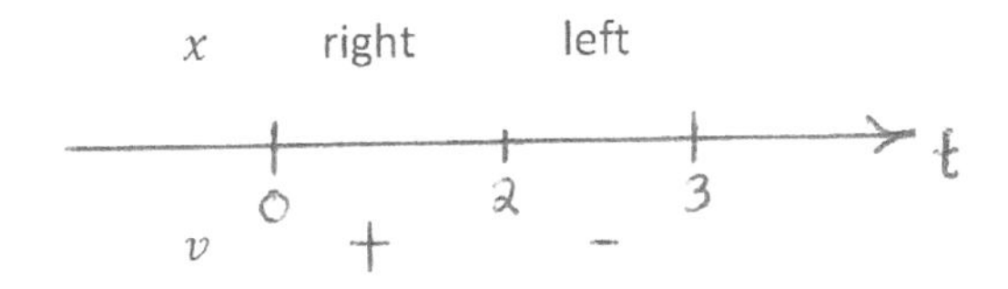

So the distance traveled = $\int_0^2 v(t)dt - \int_2^3 v(t)dt = \frac{29}{6}$

Vertical Freefall

We will consider freefall as if it were in a vacuum so that the only force acting on the object is the force of gravity. Now the acceleration due to gravity is $a(t) = -32\frac{ft}{sec^2}$ and $a = \frac{dv}{dt}$ so:

$$\frac{dv}{dt} = -32$$

$$dv = -32dt$$

$$\int dv = -\int 32dt$$

$$v = -32t + c$$

Since v depends on time t

$$v(t) = -32t + c$$

To find c set $t = 0$

$$v(0) = -32(0) + c$$

And $v(0) = c$ (i.e. c is the initial velocity)

So, $v(t) = -32t + v_0$ (where $v_0 = v(0)$)

Now, since $v = \frac{dy}{dt}$ (y because of vertical motion)

$$\frac{dy}{dt} = -32t + v_0$$

$$dy = (-32t + v_0)dt$$

$$\int dy = \int(-32t + v_0)dt$$

$$y = -16t^2 + v_0 t + c$$

Since y depends on time t

$$y(t) = -16t^2 + v_0 t + c$$

To find c set $t = 0$

$$y(0) = -16 \cdot 0^2 + v_0 \cdot 0 + c$$

And $y(0) = c$ (i.e. c is the initial height) So, $y(t) = -16t^2 + v_0 t + y_0$ (where $y_0 = y(0)$)

So starting from the acceleration due to gravity and a few basic calculus steps we were able to obtain a formula for the velocity and the position of an object that is dropped or thrown up or down as long as the initial position and velocity are known. This is only in an idealized situation. In the real world there is air resistance.

Example: A ball is thrown up at 30 ft/sec from the top of a 100 ft building

a) How long until it reaches its maximum height?

$v_0 = 30$ and $y_0 = 100$

so, $v(t) = -32t + 30$

Set $v(t) = 0$ (since that's when it stops rising) and solve for t

$0 = -32t + 30$

$t = \frac{15}{16}$ seconds

b) What is the maximum height?

We plug $\frac{15}{16}$ into the y equation which is $y(t) = -16t^2 + 30t + 100$

So, $y\left(\frac{15}{16}\right) = -16 \cdot \frac{15^2}{16} + 30 \cdot \frac{15}{16} + 100$

and $y\left(\frac{15}{16}\right) \approx 114.1$ ft

c) How long until it hits the ground?

Set $y(t) = 0$ (since that is when it's on the ground) and solve for t

We get $t \approx 3.6$ seconds

d) How fast is it going when it hits the ground?

Substitute 3.6 seconds into $v(t)$ and

$v(3.6) \approx -85.2$ ft/sec (it's negative because its falling)

DIFFERENTIAL EQUATIONS

Differential Equations

A differential equation is an equation that contains derivatives or differentials

Examples:

1) $3x^2dx + \sqrt{y}\,dy = 0$ 2) $e^x y' - \sin x = 0$

There are many different types of differential equations. One could spend a lifetime studying differential equations. We will consider only the most basic. These are called first order separable linear differential equations. The solution of any differential equation is the equation that it comes from.

Example: We see $y = \sin x$ solves the differential equation $y' - \cos x = 0$ because $\frac{d(\sin x)}{dx} = \cos x$ so $y' - \cos x = cox\ x - \cos x = 0$.

Solutions however can be difficult or impossible to find. When solving them we prefer "closed form" solutions i.e. solutions that can be written as a few simple expressions. Many solutions however are infinite series expansions or just numerical approximations. Most are unsolvable. In the eighteenth and nineteenth centuries most mathematicians were also the leading scientists of the time and the differential equations that arose in their scientific work became the focus of their mathematics. Their method of solution is often named after them.

We will turn our attention now to first order separable linear differential equations. They are called this because to solve them you "separate" variables to be with their differentials.

Example:

$$y^3y' - x^2 = 0$$

$$y^3\frac{dy}{dx} = x^2$$

$$y^3dy = x^2dx$$

$$\int y^3dy = \int x^2dx$$

$$\frac{y^4}{4} + c_1 = \frac{x^3}{3} + c_2$$

$$3y^4 - 4x^3 = c \text{ (where } c = 12(c_2 - c_1))$$

Exact Differential Equations

An exact differential equation is the total differential of some $F(x, y)$. It has the form:

$$P(x, y)\,dx + Q(x, y)\,dy = 0$$

where

$$\frac{\partial P}{\partial y} = \frac{\partial Q}{\partial x}$$

This means that the derivative of the expression $P(x, y)$ with respect to y holding x constant is equal to the derivative of the expression $Q(x, y)$ with respect to x holding y constant. If this holds true then the solution $F(x, y)$ can be determined.

In practice not many differential equations are exact in any of the natural sciences. At one time only the best mathematicians were able to provide solutions to the differential equations that arose in physics. There are many methods that these mathematicians have developed over the years since Leibniz. In fact you are required to learn many of them in a differential equations class. There are ordinary differential equations classes and partial differential equations classes. Today computers can give very good approximations to differential equations so scientists no longer have to remember every method or spend hours or days solving them.

To solve an exact differential equation means to find an expression $F(x, y)$ where $\frac{\partial F}{\partial x} = P(x, y)$ and $\frac{\partial F}{\partial y} = Q(x, y)$. We will do two examples to illustrate.

Example 1: $(2xy)dx + (x^2)dy = 0$ is exact since $\frac{\partial(2xy)}{\partial y} = 2x = \frac{\partial(x^2)}{\partial x}$

To solve this differential equation we can integrate $P(x, y)$ treating y as a constant:

$\int 2xydx = x^2y + g(y)$, the $g(y)$ replaces the "constant" c since y is treated as a constant.

Now we differentiate $x^2y + g(y)$ with respect to y holding x constant (partial derivative).

$\frac{\partial(x^2y + g(y))}{\partial y} = x^2 + g'(y)$. Now note that this is $Q(x, y)$ with $g'(y) = 0$.

So the solution of the differential equation must be: $f(x, y) = x^2y + c$

Example 2: $(\cos y)dx + (y^2 - x \sin y)\, dy = 0$ is exact since $\frac{\partial(\cos y)}{\partial y}$ = - $\sin y$ = $\frac{\partial(y^2 - x\sin y)}{\partial x}$ Proceeding as we did on the previous example:

To solve this differential equation we again integrate $P(x, y)$ treating y as a constant.

$\int \cos y dx = x\cos y + g(y)$ the $g(y)$ replaces "constant" c since y is treated as a constant

Now we differentiate $x\cos y + g(y)$ with respect to y holding x constant (partial derivative).

$\frac{\partial(x\cos y + g(y))}{\partial y}$ = - $x \sin y + g\,'(y)$ Now, since this must be $Q(x, y)$ then $g\,'(y) = y^2$.

To find $g(y)$ we integrate normally. Since $g\,'(y) = y^2$ that means that:

$$\frac{d(g(y))}{dy} = y^2$$

$$d(g(y)) = y^2\, dy$$

$$\int d(g(y)) = \int y^2 dy$$

$$g(y) = \frac{y^3}{3} + c$$

So the solution must be: $F(x, y) = \frac{y^3}{3} - x \sin y + c$

GROWTH/DECAY

Exponential Growth and Decay

If the rate that a population P grows is directly proportional to the population P present then we have a differential equation of the form:

$$\frac{dP}{dt} = kP$$

This is a separable differential equation. Solving for P where t is time we get:

$$dP = kPdt$$

$$\frac{dP}{P} = kdt$$

$$\int \frac{dP}{P} = \int kdt$$

$$\ln|P| + c_1 = kt + c_2$$

$$\ln P = kt + c_3 \quad (c_3 = c_2 - c_1)$$

$$e^{\ln P} = e^{(kt + c_3)}$$

$$P = e^{kt} \cdot e^{c_3}$$

$P = e^{kt} \cdot c$ (where $c = e^{c_3}$ which is a constant)

$$P = ce^{kt}$$

P depends on t i.e. $P(t)$ so,

$$P(t) = ce^{kt}$$

Now set $t = 0$ and solve for c:

$$P(0) = ce^{k \cdot 0}$$

So $P(0) = c$ is the initial population

And $P(t) = P(0)e^{kt}$

We have derived an equation that gives the population P at any given time t.

The expression e^k is called the growth/decay factor. If k is positive then it's growth, if k is negative then it's decay. Also $(e^k - 1) \cdot 100$ is the growth/decay rate (as a percent).

Examples:

1) If $P(t) = 500e^{.299t}$ then what is the growth rate?

$$(e^{.299} - 1) \cdot 100 \approx 34.9\%$$

2) If the growth rate of a population is 15.3% then find k.

$$\text{Since } (e^k - 1) \cdot 100 = 15.3$$

$$e^k - 1 = .153$$

$$e^k = 1.153$$

$$\ln e^k = \ln 1.153$$

$$k \approx .14$$

3) If an insect population is given by $P(t) = 650e^{.03t}$ (t measured in weeks) then how many will there be in 12 weeks?

Set $t = 12$ and the population will be $P(12) = 650e^{.03 \cdot 12} \approx 932$ insects.

4) When will there will be 2000 insects?

Set $P(t) = 2000$ and solve for t

$$2000 = 650e^{.03t}$$

$$\frac{40}{13} = e^{.03t}$$

$$\ln \frac{40}{13} = \ln e^{.03t}$$

$$\ln \frac{40}{13} = .03t$$

$$t \approx 37.5 \text{ weeks}$$

We said previously that the growth/decay factor of the expression $P(t) = P(0)e^{kt}$ is e^k. In some situations we are given the growth factor. For instance we may be told that a population is doubling in a unit of time or we may be given the half-life of a radioactive substance in a unit of time. In problems like these we can replace e^k with that growth/decay factor we will label as B. So the above population equation becomes

$$P(t) = P(0)B^t$$

Example 1) Write the population equation if a population of frogs triples every year with an initial population of 25 frogs.

The population equation is $P(t) = 25 \cdot 3^t$

If it takes more or less than a unit of time for the growth/decay factor to occur then this time factor t_t is the denominator of t. The new equation becomes

$$P(t) = P(0)B^{t/t_t}$$

Example 2) A population starts with 500 and doubles every 5 years. Write the population equation.

The equation is $P(t) = 500 \cdot 2^{t/5}$

If a population decreases then B is a proper fraction.

Example 3) The half-life of a certain radioactive substance is 1000 years with 10 grams to start with.

a) Write the amount present at any given time.

The amount present at any given time is $A(t) = 10\left(\frac{1}{2}\right)^{t/1000}$

b) How many grams will there be in 300 years?

Set $t = 300 \Rightarrow A(300) = 10\left(\frac{1}{2}\right)^{300/1000} \approx 8.1$ grams

c) How long until there is 4 grams?

Set $A(t) = 4$ and solve for t

$$4 = 10\left(\frac{1}{2}\right)^{t/1000}$$

$$4/10 = \left(\frac{1}{2}\right)^{t/1000}$$

$$log_{1/2}\, 2/5 = log_{1/2} \left(\frac{1}{2}\right)^{t/1000}$$

$$log_{1/2}\, 2/5 = t/1000$$

$$t \approx 1{,}322 \text{ years}$$

Example 4) The rate of a chemical reaction is proportional to the amount present. If the rate is $\frac{dy}{dt}$ = -0.2 y and there are 50 grams to start while the change takes place in hours then how many grams will there be in 2 hours?

The solution of this differential equation is $y = y(0)e^{kt}$ where k = -0.2 so,

$$y = 50e^{-0.2t}$$

In 2 hours there will be: $y(2) = 50e^{-0.2 \cdot 2}$

$$y(2) \approx 33.5 \text{ grams}$$

Newton's Law of Cooling

Newton's law of cooling says that the rate of change of the temperature of an object that is cooling is directly proportional to the difference between it and its surrounding temperature. This is a separable differential equation:

$\frac{dT}{dt} = k(T - T_s)$ where T is objects initial temperature and T_s is surrounding temperature.

So separate variables and solve for T:

$$\frac{dT}{(T-T_s)} = kdt$$

$$\int \frac{dT}{(T-T_s)} = \int kdt$$

$$\ln(T - T_s) + c_1 = kt + c_2$$

$$\ln(T - T_s) = kt + c_3 \text{ (where } c_3 = c_2 - c_1\text{)}$$

$$e^{\ln(T-T_s)} = e^{(kt+c_3)}$$

$$T - T_s = e^{kt} \cdot e^{c_3}$$

$$T - T_s = ce^{kt} \quad \text{(where } c = e^{c_3} \text{ which is a constant)}$$

$$T = ce^{kt} + T_s$$

Now T depends on time t i.e. $T(t)$ so,

$$T(t) = ce^{kt} + T_s$$

Now set $t = 0$ and solve for c:

$$T(0) = ce^{k \cdot 0} + T_s$$

$$T(0) = c + T_s$$

$$T(0) - T_s = c$$

So $T(t) = (T(0) - T_s)e^{kt} + T_s$

And $T(t) = (T_0 - T_s)e^{kt} + T_s$ (where $T_0 = T(0)$)

From Newton's law of cooling we have a formula for the objects temperature T at any time t.

Example: Suppose a cup of coffee cools from 100^0 to 60^0 in 20 minutes where the room is 40^0. How much longer will it take to cool to 50^0?

First solve for k:

$$T(t) = (T_0 - T_s)e^{kt} + T_s$$

$$60 = (100 - 40)e^{20k} + 40$$

$$20 = 60e^{20k}$$

$$\frac{1}{3} = e^{20k}$$

$$ln\frac{1}{3} = ln\, e^{20k}$$

$$ln\frac{1}{3} = 20k$$

$$k = \frac{ln\frac{1}{3}}{20} \approx -.055$$

So $T(t) = 60e^{-.055t} + 40$

To solve the question asked, how much longer will it take to cool to 50^0?

Set $T(t)$ = 50 and solve for t

$$50 = 60e^{-.055t} + 40$$

$$10 = 60e^{-.055t}$$

$$\frac{1}{6} = e^{-.055t}$$

$$ln\frac{1}{6} = ln\, e^{-.055t}$$

$$ln\frac{1}{6} = -.055t$$

So $t = \frac{ln\frac{1}{6}}{-.055} \approx$ 32.6 minutes total or 12.6 minutes longer.

Logistic Growth

The exponential growth model that we used previously applies only in "ideal" conditions i.e. no constraints on the growth. Unbounded growth however is realistically impossible. Real world growth may be exponential to start with but other factors will enter that will inhibit this continued type of growth. One such factor is the "saturation level." Sooner or later the environment will be saturated further growth becomes impossible and the exponential curve will flatten.

In the 1830's a Belgian mathematician P.F. Verhulst provided us with the logistic equation which takes account of saturation. The differential equation is:

$$\frac{dP}{dt} = kP\left(1 - \frac{P}{L}\right)$$

Where P = population

k = the constant of proportionality

L = the carrying capacity of the environment

The solution to this differential equation is:

$$P(t) = \frac{L}{1+ce^{-kt}}$$

Where $P(0) = P_0 = \frac{L}{1+c}$

And $c = \frac{L - P_0}{P_0}$

The logistic Curve

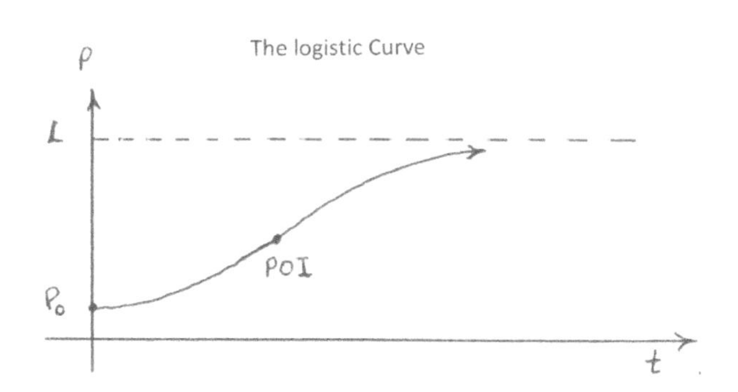

Example: If the rate of growth of a fish population in a lake is $\frac{dP}{dt} = \frac{P}{10}\left(1 - \frac{P}{5000}\right)$ where P_0 = 500 fish and time is measured in months.

a) Find the population equation.

$$k = \frac{1}{10} \quad c = \frac{5000 - 500}{500} = 9$$

$$P(t) = \frac{5000}{1+9e^{-.1t}}$$

b) How many fish will there be in 15 months?

$$P(15) \approx \frac{5000}{1+9e^{-.1 \cdot 15}} = 1{,}662 \text{ fish}$$

c) How long until there are 4,000 fish?

Set $P(t)$ = 4000 and solve for t

$$4000 = \frac{5000}{1+9e^{-.1t}}$$

$$4000(1 + 9e^{-.1t}) = 5000$$

$$1 + 9e^{-.1t} = \frac{5}{4}$$

$$9e^{-.1t} = \frac{1}{4}$$

$$e^{-.1t} = \frac{1}{36}$$

$$-.1t = ln\frac{1}{36}$$

$$t \approx 35.8 \text{ months}$$

The reverse problem: Given that $P(t) = \frac{500}{1+4e^{-t/2}}$ Find $\frac{dP}{dt}$

$$L = 500 \quad k = \frac{1}{2} \quad \text{So} \quad \frac{dP}{dt} = \frac{P}{2}\left(1 - \frac{P}{5000}\right)$$

INVERSE FUNCTIONS

Derivatives of Inverses

In most standard calculus textbooks this topic is probably one of the most difficult. All texts of course discuss the inverse trig and log functions and derive formulas for their derivatives. We will do the same in the next chapter using non-rigorous infinitesimal methods. But the derivation of the derivative of an inverse in general is very difficult to follow when it's dressed in all of the "limit garb." The result is usually presented as;

$$D(f^{-1}(x)) = \frac{1}{D(f(x))}$$

with the rejoinder that the x in f^{-1} is the y of f. When I first took calculus the derivation of this expression may as well have been written in Chinese! Of course, I was eventually able to make sense of it but I discovered that teaching it was more of a nightmare than learning it. Using differentials, the formula actually makes sense and a derivation seems almost redundant:

$$\frac{dx}{dy} = \frac{1}{\frac{dy}{dx}}$$

Recall from a previous discussion that $\frac{dx}{dy}$ implies that x depends on y i.e. $x(y)$. So $\frac{dx}{dy}$ is the derivative of $x(y)$, the inverse of $y(x)$. The right side of the equation implies that the reciprocal of $\frac{dy}{dx}$ is used to determine it.

Example 1) Let y = x^2 Given that y(3) = 9 and $x(9) = 3$ find $\frac{dx}{dy}|_{y=9}$

$$\text{We know that } \frac{dy}{dx} = \frac{d(x^2)}{dx} = 2x$$

$$\text{So, } \frac{dx}{dy}|_{y=9} = \frac{1}{\frac{dy}{dx}}|_{x=3} = \frac{1}{2x}|_{x=3} = \frac{1}{6}$$

We can also find the inverse formula of y = x^2 which is $x = \sqrt{y}$ then its derivative, which is $\frac{dx}{dy} = \frac{1}{2\sqrt{y}}$ and evaluate it at y = 9 which is $\frac{dx}{dy}|_{y=9} = \frac{1}{2\sqrt{y}}|_{y=9} = \frac{1}{2\sqrt{9}} = \frac{1}{6}$

The derivative of the inverse can also be written as:

$$x'(y) = \frac{1}{y'(x)}$$

If we want the derivative of the inverse of a complicated expression that we can't find the inverse formula for, it's good to know that we don't have to find it and we can still get its derivative.

Example 2) Let y = $2x^5 - x^4 + 3x^3 + x^2 - x + 1$. Given that y(1) = 5, find $x\,'(5)$

$$y\,' = 10x^4 - 4x^3 + 9x^2 + 2x - 1 \text{ and } y\,'(1) = 16 \text{ so,}$$

$$x\,'(5) = \frac{1}{y\,'(1)} = \frac{1}{16}$$

Let's consider a trig function now:

Example 3) Let y = $\sin x$. Given that y(π/3) = √3/2, find $x\,'(\sqrt{3}/2)$

$$y\,' = \cos x \text{ and } \cos(\pi/3) = \frac{1}{2} \text{ so,}$$

$$x\,'(\sqrt{3}/2) = \frac{1}{y\,'(\pi/3)} = \frac{1}{\cos(\pi/3)} = 2$$

Turning now to an exponential function:

Example 4) Let y = e^x Given that y(0) = 1, find $x\,'(1)$

$$\text{Now, } y\,' = e^x \text{ so,}$$

$$x\,'(1) = \frac{1}{y\,'(0)} = \frac{1}{e^0} = \frac{1}{1} = 1$$

The reader can easily check the derivatives of the actual inverse formulas of the two functions in examples 3 and 4 to see their correctness.

Parametric Equations

If a particle is moving linearly in the xy plane as a function of time then the standard form of a line won't be of much help because the variables involve only x and y, no variable for time. In this case parametric equations of a line would be used instead. Suppose the x coordinate of the particle was given by $x(t) = 2 + 4t$ and the y coordinate by $y(t) = 5 + 3t$ where t represents time. What is the initial position of the particle? Set $t = 0$ and $x(0) = 2$ and $y(0) = 5$. So the particle starts at (2, 5). If you make a table of t values and plot the points you will see a line. We call these equations parametric equations of the line. The general form is:

$$x(t) = x_1 + m_x t$$

$$y(t) = y_1 + m_y t$$

Here, when $t = 0$ (x_1, y_1) would be the initial value or starting point and $\frac{m_y}{m_x} = m$ the slope of the line. Recall that we said (page 10) $x(t)$ just means that x is a function of t, so we can write:

$$x = x_1 + m_x t$$

$$y = y_1 + m_y t$$

Example 1) Let's use the parametric equations from above, $x = 2 + 4t$ $y = 5 + 3t$. A good graphing calculator can graph these parametric equations. Note that if you plug in any time value, you can determine the coordinates of the moving particle. To convert these equations to a Cartesian point-slope equation, just use the initial point (2, 5) and the slope $\frac{m_y}{m_x} = \frac{3}{4}$ and we get the equation, $y - 5 = \frac{3}{4}(x - 2)$ Note that this equation can't provide the position for given time values.

If we start with a Cartesian equation of a line like $y = \frac{2}{3}x + 4$, we can use (0, 4) as the initial point and $2 = m_y$, $3 = m_x$ and write $x = 0 + 4t$ and $y = 4 + 3t$. But a much easier way is simply let $x = t$ and write $x = t$ and $y = 4 + 3t$. The nice thing about this method is that it works for any Cartesian equations not just lines.

Example2) Let $y = x^2 - x + 1$. The parametric equations for this parabola are:

$$x = t$$

$$y = t^2 - t + 1$$

Derivatives of Parametric Equations

If a curve has been defined parametrically i.e. $x = x(t)$ and $y = y(t)$ and a particle is moving along the curve then the rate of x is $\frac{dx}{dt} = x'(t)$ and the rate of y is $\frac{dy}{dt} = y'(t)$. We could also say that the velocity in the x direction is $x'(t)$ and in the y direction $y'(t)$. The slope of the curve itself we know is $\frac{dy}{dx} = y'(x)$ but this easily follows since:

$$\frac{\frac{dy}{dt}}{\frac{dx}{dt}} = \frac{dy}{dt} \cdot \frac{dt}{dx} = \frac{dy}{dx} = y'(x)$$ So, the slope of the curve at x is

$$y'(x) = \frac{y'(t)}{x'(t)}$$

Example 1) : Find $y'(x)$ for the parametric equations; $x = t^2 - 1$ and $y = t^3 + 3$

Since $x'(t) = 2t$ and $y'(t) = 3t^2$ So,

$$y'(x) = \frac{y'(t)}{x'(t)} = \frac{3t^2}{2t} = \frac{3t}{2}$$ Note it's actually expressed in t

Second Derivatives of Parametric Equations

The second derivative $\frac{d^2y}{dx^2}$ is used to determine concavity. For parametric equations the second derivative is:

$$\frac{d^2y}{dx^2} = \frac{d}{dx}\left(\frac{dy}{dx}\right) = \frac{d(y')}{dx} \cdot \frac{\frac{1}{dt}}{\frac{1}{dt}} = \frac{\frac{d(y')}{dt}}{\frac{dx}{dt}}$$ Note the multiplication by $\frac{\frac{1}{dt}}{\frac{1}{dt}} = 1$

Note, as in the example above, y' will already be expressed in terms of t so the numerator is just the next derivative of y' with respect to t over the derivative of x with respect to t.

Example 2): Given that $x = \frac{t^2}{2}$ and $y = t^4$ find the second derivative $\frac{d^2y}{dx^2}$

$$\frac{dx}{dt} = t \text{ and } \frac{dy}{dt} = 4t^3 \text{ so, } y' = \frac{dy}{dx} = \frac{\frac{dy}{dt}}{\frac{dx}{dt}} = \frac{4t^3}{t} = 4t^2 \quad \Rightarrow \quad \frac{d^2y}{dx^2} = \frac{\frac{dy'}{dt}}{\frac{dx}{dt}} = \frac{8t}{t} = 8$$

Since parametric equations can be used to generate curves that are not functions and are usually studied with polar coordinates, we will omit the integral for parametric equations.

Vectors

If a particle is moving in the xy plane with magnitude and direction, its motion can be represented as a vector. The topic of vectors is too rich to do justice with a cursory discussion of them in this little volume dedicated to infinitesimal calculus. I refer the reader to either online resources or your own textbook on calculus. But a good grounding in vectors is essential in science. We will simply state that a vector **r**(t) written in component form is:

$$\mathbf{r}(t) = x(t)\,\hat{\boldsymbol{\imath}} + y(t)\,\hat{\boldsymbol{\jmath}}$$

where $\hat{\boldsymbol{\imath}}$ and $\hat{\boldsymbol{\jmath}}$ are the standard unit normal vectors along the x, y axes. These are sometimes written as $\hat{\boldsymbol{x}}$ and $\hat{\boldsymbol{y}}$.

Differentiating Vectors

Differentiating is direct and straight forward:

$$d(\mathbf{r}(t)) = d(x(t))\,\hat{\boldsymbol{\imath}} + d(y(t))\,\hat{\boldsymbol{\jmath}}$$

$$\frac{d(\mathbf{r}(t))}{dt} = \frac{d(x(t))}{dt}\hat{\boldsymbol{\imath}} + \frac{d(y(t))}{dt}\hat{\boldsymbol{\jmath}}$$

$$\mathbf{r}'(t) = x'(t)\,\hat{\boldsymbol{\imath}} + y'(t)\,\hat{\boldsymbol{\jmath}}$$

Example 1) Suppose $\mathbf{r}(t) = sin(t)\,\hat{\boldsymbol{\imath}} + t^2\hat{\boldsymbol{\jmath}}$ Find $\mathbf{r}'(t)$

In this case $x(t) = sin(t)$ and $y(t) = t^2$ So, $x'(t) = cos(t)$ and $y'(t) = 2t$ so,

$$\mathbf{r}'(t) = cos(t)\,\hat{\boldsymbol{\imath}} + 2t\,\hat{\boldsymbol{\jmath}}$$

If t is time then $\mathbf{r}'(t) = \mathbf{v}(t)$ i.e. the velocity of the particle at time t. To find the second derivative we simply take the derivative of the first derivative. This will give acceleration vector.

$$\mathbf{r}''(t) = x''(t)\,\hat{\boldsymbol{\imath}} + y''(t)\,\hat{\boldsymbol{\jmath}} = \mathbf{a}(t)$$

Example 2) Find $\mathbf{r}''(t)$ for the previous example

$x'(t) = cos(t)$ so, $x''(t) = -sin\,t$ and $y'(t) = 2t$ so $y''(t) = 2$ therefore

$$\mathbf{r}''(t) = -sin(t)\,\hat{\boldsymbol{\imath}} + 2\hat{\boldsymbol{\jmath}}$$

Example 3) Let $\mathbf{r}(t) = (t^2 - 4)\,\hat{\boldsymbol{\imath}} + (t)\,\hat{\boldsymbol{\jmath}}$ find **v**(t) and **a**(t) at $t = 1$ then graph

$$\mathbf{v}(t) = 2t\hat{\boldsymbol{\imath}} + 1 \text{ and } \mathbf{a}(t) = 2\hat{\boldsymbol{\imath}}$$

$$\mathbf{v}(1) = 2\hat{\imath} + 1\hat{\jmath} \quad \mathbf{a}(1) = 2\hat{\imath} \quad \text{also } \mathbf{r}(1) = -3\hat{\imath} + 1\hat{\jmath}$$

The graph of **r** is pictured below.

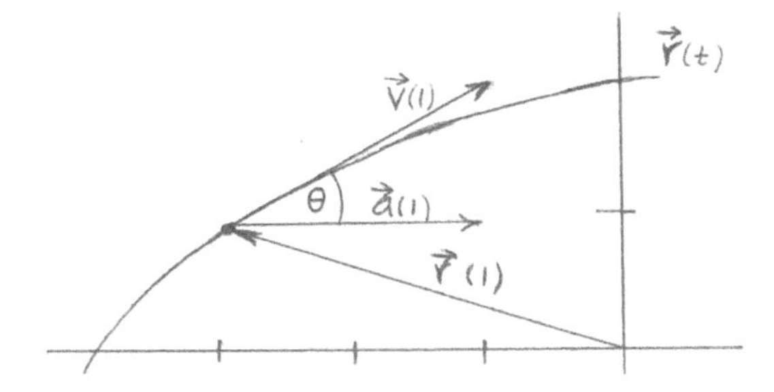

If we place the tails of **v** and **a** on the head of **r** we get the vectors pictured on the left

The angle between two vectors is given by: $\theta = cos^{-1}\left(\frac{\mathbf{a}\bullet\mathbf{b}}{\|\mathbf{a}\|\cdot\|\mathbf{b}\|}\right)$. Using this formula we can determine the angle between the velocity and acceleration vectors.

If the angle between $\mathbf{v}(t)$ and $\mathbf{a}(t)$ is acute, right, or obtuse then:

1) If the angle is acute then the particle is speeding up.

2) If the angle is a right angle then the particle is moving at a constant speed

3) If the angle is obtuse then the particle is speeding up

To find this angle we will need three formulas; the angle formula above, the dot product, and the magnitude of a vector formula. The dot product of two vectors $\mathbf{a} = x_1(t)\,\hat{\imath} + y_1(t)\,\hat{\jmath}$ and $\mathbf{b} = x_2(t)\,\hat{\imath} + y_2(t)\,\hat{\jmath}$ is:

$$x_1(t) \cdot x_2(t) + y_1(t) \cdot y_2(t)$$

The magnitude of vector formula is: $\|\ \mathbf{a}\ \| = \sqrt{\mathrm{x}(t)^2 + \mathrm{y}(t)^2}$

Let's find the angle between the velocity and acceleration vectors in the example above:

$$cos^{-1}\left(\frac{2\cdot2+1\cdot0}{\sqrt{2^2+1^2}\cdot\sqrt{2^2+0^2}}\right) = cos^{-1}\left(\frac{4}{2\sqrt{5}}\right) = 26.6 \text{ degrees}$$

This angle is acute so this means that the particle is speeding up.

Lastly, the speed of the particle is the magnitude of the velocity $\|\ \mathbf{v}\ \| = \sqrt{\mathrm{x}'(t)^2 + \mathrm{y}'(t)^2}$ so the actual speed of the particle in the previous example is $\sqrt{2^2 + 1^2} = \sqrt{5}$

DERIVATIONS

Infinitesimal Methods of Deriving Formulas

In order to derive the formulas we have been using we will use three infinitesimal properties:

1) dx^2 and higher powers of dx are infinitely smaller than dx. We will therefore consider their value zero.

2) In any expression containing infinitesimals of different orders only the lowest order are retained.

3) An infinitesimal can be added to or subtracted from any real number without augmenting or diminishing that real number.

Notation

If x changes by the infinitesimal amount dx to become $x + dx$ and y changes by the infinitesimal amount dy to become $y + dy$, then we can use the following expression to find dy.

$$y + dy = y(x + dx)$$

$$dy = y(x + dx) - y(x)$$

We will use this expression to make most of our derivations.

Formulas

Recall that $d(variable)$ represents the infinitesimal change that occurs in that variable during an instant of time dt so the differential of a variable u is du i.e. clearly formula 1.

1) $d(u) = du$

Now, constants like 5 don't change, they're constant. The differential of a constant is therefore zero. This brings us to formula 2, the differential of a constant c.

2) $d(c) = 0$ (where c is a constant). This holds because constants don't change.

3) $d(x^2) = 2xdx$

Let $y = x^2$

$y + dy = (x + dx)^2$

$dy = x^2 + 2xdx + dx^2 - y(x)$

$dy = x^2 + 2xdx + dx^2 - x^2$

$dy = 2xdx + dx^2$

$dy = 2xdx$

$d(x^2) = 2xdx$

4) $d(x^3) = 3x^2dx$

Let $y = x^3$

$y + dy = (x + \mathrm{dx})^3$

$dy = x^3 + 3x^2dx + 3xdx^2 + dx^3 - y(x)$

$dy = x^3 + 3x^2dx + 3xdx^2 + dx^3 - x^3$

$dy = 3x^2dx + 3xdx^2 + dx^3$

$dy = 3x^2dx$

$d(x^3) = 3x^2dx$

5) $d(x^n) = nx^{n-1}dx$ This is a generalization of 3 and 4 The derivation involves the binomial

Expansion. We will omit it.

General Formulas

Let u and v be dependent or independent variables. The differentials of their sum, difference, product and quotient are determined as follows:

1) $d(u+v) = du + dv$

Let $y = u + v$

$y + dy = u + du + v + dv$

$u + v + dy = u + v + du + dv$

$dy = du + dv$

$d(u+v) = du + dv$

2) Subtraction is done similarly so $d(u-v) = du - dv$

3) $d(uv) = udv + vdu$

Let $y = uv$

$y + dy = (u + du)(v + dv)$

$y + dy = uv + udv + vdu + dudv$

$uv + dy = uv + udv + vdu + dudv$

$dy = udv + vdu + dudv$

$dy = udv + vdu$

$d(uv) = udv + vdu$

4) $d\left(\frac{u}{v}\right) = \frac{vdu-udv}{v^2}$

Let $y = \left(\frac{u}{v}\right)$

$y + dy = \frac{u+du}{v+dv}$

$y + dy = \frac{(u+du)(v-dv)}{(v+dv)(v-dv)}$

$y + dy = \frac{uv-udv+vdu+dudv}{v^2-vdv+vdv-dv^2}$

$y + dy = \frac{uv-udv+vdu}{v^2}$

$\frac{u}{v} + dy = \frac{u}{v} - \frac{udv}{v^2} + \frac{vdu}{v^2}$

$dy = \frac{vdu}{v^2} - \frac{udv}{v^2}$

$d\left(\frac{u}{v}\right) = \frac{vdu-udv}{v^2}$

5) $d(cu) = cdu$

Let y = cu

$dy = d(cu) = cdu + udc$ But c is constant so $dc = 0$

$d(cu) = cdu$ Constants can be moved across a differential NOT variables.

Trig Functions

Madhava of Sangamagrama (1350 – 1425) in India, founder of the Kerala School of Astronomy and Mathematics showed that $sin\,x$ can be expanded into an infinite series of the form:

$$sin\,x = x - \frac{x^3}{3!} + \frac{x^5}{5!} - \frac{x^7}{7!} + \frac{x^9}{9!} \ . \ . \ .$$

Newton and Leibnitz also derived this same series in Europe many years later. Since

$$sin\,x = x - \frac{x^3}{3!} + \frac{x^5}{5!} - \frac{x^7}{7!} + \frac{x^9}{9!} \;.\;.\;.\; \text{we have:}$$

$$sin\,dx = dx - \frac{dx^3}{3!} + \frac{dx^5}{5!} - \frac{dx^7}{7!} + \frac{dx^9}{9!} \;.\;.\;.\; \text{so:}$$

$$sin\,dx = dx$$

Note also that from a trig identity, $cos\,dx = \sqrt{1-(sindx)^2} = \sqrt{1-(dx)^2} = \sqrt{1} = 1$

So: $sin\,dx = dx$ and $cos\,dx = 1$

We need these facts to derive the next formulas.

1) $d(sin\,x) = cos\,x\,dx$

Let $y = sin\,x$

$y + dy = sin(x + dx)$

$y + dy = sin\,x\,cos\,dx + cos\,x\,sin\,dx$

$y + dy = sin\,x \cdot 1 + cos\,x \cdot dx$

$sin\,x + dy = sin\,x + cos\,x\,dx$

$dy = cos\,xdx$

$d(\sin x) = cos\,x\,dx$

2) $d(cos\,x) = -\,sin\,x\,dx$ The derivation is similar to 1) and is left to the reader

3) $d(tan\,x) = (sec\,x)^2\,dx$

$$d(tan\,x) = d\left(\frac{sin\,x}{cos\,x}\right) = \frac{cos\,xd(sin\,x) - sin\,xd(cos\,x)}{(cos\,x)^2} = \frac{cos\,x\,cos\,xdx - sin\,x(-\,sin\,x)dx}{(cos\,x)^2}$$

$$= \frac{(cos\,x)^2dx}{(cos\,x)^2} + \frac{(sin\,x)^2dx}{(cos\,x)^2} = dx + (tan\,x)^2\,dx = (1 + (tan\,x)^2)dx = (sec\,x)^2\,dx$$

The remaining three trig differentials are all derived using the quotient rule applied to their reciprocals and are left to the reader.

Inverse Trig functions

To derive the differentials we can make use of the diagram below and some basic substitutions.

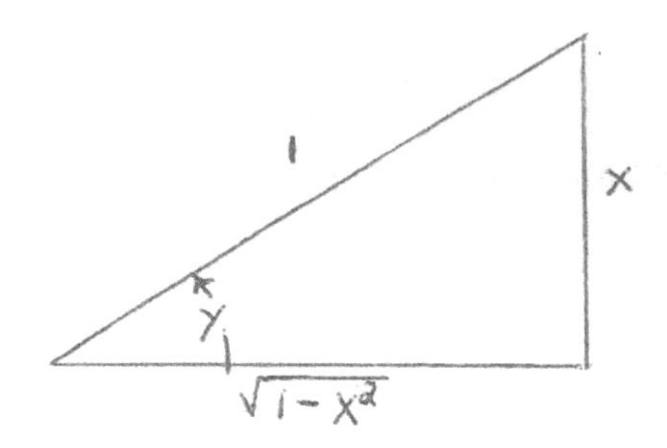

1) $d(sin^{-1} x) = \frac{dx}{\sqrt{1-x^2}}$

Let y = $sin^{-1} x$

$sin\, y = x$

$d(sin\, y) = dx$

$cos\, ydy = dx$

$dy = \frac{dx}{cos\, y}$

$d(sin^{-1} x) = \frac{dx}{\sqrt{1-x^2}}$

The derivations of the $cos^{-1} x,\ tan^{-1} x$ are proved similarly so they are left to the reader.

The Natural Log Function

In 1668 the mathematician N. Mercator showed that:

$ln(1+x) = x - \frac{x^2}{2} + \frac{x^3}{3} - \frac{x^4}{4} + \frac{x^5}{5} \ . \ . \ .$ Using this we can show that $d(ln\,x) = \frac{dx}{x}$

1) Let $y = ln\,x$

$$y + dy = ln(x + dx)$$

$$dy = ln(x + dx) - y$$

$$dy = ln(x + dx) - ln\,x$$

$$dy = ln\left(\frac{x + dx}{x}\right)$$

$$dy = ln\left(1 + \frac{dx}{x}\right)$$

$$dy = \frac{dx}{x} - \frac{dx^2}{2x^2} + \frac{dx^3}{3x^3} - \frac{dx^4}{4x^4} + \frac{dx^5}{5x^5} \ . \ . \ .$$

$$dy = \frac{dx}{x}$$

$$d(ln\,x) = \frac{dx}{x}$$

It is easy now to show that $d(e^x) = e^x dx$

2) Let $y = e^x$

$$ln\,y = ln\,e^x$$

$$ln\,y = x$$

$$d\,(ln\,y) = dx$$

$$\frac{dy}{y} = dx$$

$$dy = ydx$$

$$d(e^x) = e^x dx$$

Differentials of a^x and $log_a x$

1) $d(log_a x) = d\left(\frac{ln\,x}{ln\,a}\right)$

$d\left(\frac{ln\,x}{ln\,a}\right) = \frac{1}{ln\,a} \cdot d(ln\,x)$

$d\left(\frac{ln\,x}{ln\,a}\right) = \frac{1}{ln\,a} \cdot \frac{dx}{x}$

$d\left(\frac{ln\,x}{ln\,a}\right) = \frac{dx}{x \cdot ln\,a}$

$d(log_a x) = \frac{dx}{x \cdot ln\,a}$

2) $d(a^x) = a^x\,ln\,a\,dx$

Let y = a^x

$log_a y = log_a a^x$

$log_a y = x$

$d(log_a y) = dx$

$\frac{dy}{x \cdot ln\,a} = dx$

$dy = x \cdot ln\,a \cdot\ dx$

$d(a^x) = a^x\,ln\,a\,dx$

Conclusion

It should be clear that the concept of the infinitesimal differential is much easier to work with than the derivative defined as a limit. The problem with the infinitesimal concept was that neither Leibniz nor the mathematicians following him were able to provide a justification for it until 1960. Meanwhile mathematicians had completed the theory of limits by the 1850's. So limit calculus became the "standard calculus" since then. Everyone has since been taught calculus with limits.

The time is long overdue to begin teaching, at least to science and engineering majors, the infinitesimal methods that are so much easier and have been vindicated by modern logic. There should be room in the Math Department for this type of course.

As mentioned previously this book is by no means a mathematical treatise or textbook. It is a purely practical exposition of the methods that were used with so much success before the theory of limits supplanted infinitesimal calculus. It is meant for the practical worker in need of simpler concepts with which to understand and apply calculus.

Much of the material discussed was presented in a purely introductory way intended to familiarize students with some of the advanced concepts they usually never see in their beginning classes. This is because most mathematics textbooks and instructors will not present it without proofs which are well beyond elementary calculus. But the concepts are not that difficult to grasp. Furthermore, the ideas are so useful in science that early exposure to these concepts will only facilitate a deeper understanding when the time is right. I have therefore included them in a simplified form for students to ponder and perhaps later pursue either online, where there are many excellent videos, or in further coursework.

Made in United States
Cleveland, OH
06 December 2024